BRITISH AIRCRAFT OF WORLD WAR ONE

A PHOTOGRAPHIC GUIDE TO MODERN SURVIVORS, REPLICAS AND REPRODUCTIONS

LEE CHAPMAN

Title page image: The sun rises behind a Royal Aircraft Factory SE5a.

Contents page image: The Great War Display Team recreates a dogfight, trying to avoid the flak.

Author's Note

The author has tried to document as many surviving World War One aircraft types as possible, with an emphasis on the types that played significant roles in the conflict. The focus of this book has been centred around aircraft that can still be seen today in museums or flying at airshows. Sadly, the aircraft built during this time were fragile, largely wooden, lightweight structures that were not built to last. Additionally, preserving aircraft was not common practice at the time and, as such, genuine complete survivors are rare. The author has deliberately included images of replicas and reproductions to illustrate what the aircraft types may have looked like and to show where and how these sites can be seen today. Some significant types used during the war are now completely extinct, and, as such, these aircraft are not fully discussed in this book.

Fortunately, some aircraft have survived and many of these survivors have been authentically restored to represent the types of aircraft as they were during World War One. Modern interest in World War One aviation has also led to an ever-increasing number of new-build aircraft that allow us to get a glimpse of what some of these aircraft would have been like over 100 years ago.

For the purposes of this book, the terms 'original', 'replica' and 'reproduction' have been used to best describe an aircraft's authenticity. An 'original' aircraft has been deemed by the author to contain a substantial number of original period parts, even though it may have been significantly restored. 'Reproduction' aircraft have been deemed as new-build aircraft built from original designs using similar materials and production methods wherever practical. A 'replica' label has been given to lookalike aircraft that appear similar on the surface but incorporate modern materials and designs. The author has used the best available knowledge to classify aircraft and recognises that this may be an over-simplification in some cases.

Lee Chapman, 2021
(www.facebook.com/ChappersPhotography)

Published by Key Books
An imprint of Key Publishing Ltd
PO Box 100
Stamford
Lincs PE19 1XQ

www.keypublishing.com

The right of Lee Chapman to be identified as the author of this book has been asserted in accordance with the Copyright, Designs and Patents Act 1988 Sections 77 and 78.

Copyright © Lee Chapman, 2021

ISBN 978 1 80282 000 3

All rights reserved. Reproduction in whole or in part in any form whatsoever or by any means is strictly prohibited without the prior permission of the Publisher.

Typeset by SJmagic DESIGN SERVICES, India.

CONTENTS

Foreword	4
Preface	6
Chapter 1 – Introduction	8
Chapter 2 – Early Aviation in Britain: Pre-war Aircraft	16
Chapter 3 – Eyes in the Sky: Reconnaissance Aircraft	26
Chapter 4 – The Birth of the Scout: Early Single-Seaters	40
Chapter 5 – An Extra Pair of Eyes: Two-Seater Aircraft	58
Chapter 6 – The Air War Matures: Advanced Single-Seaters	68
Chapter 7 – The Mount of Aces: Royal Aircraft Factory SE5a	82
Chapter 8 – Taking the Fight to the Enemy: Bombing Aircraft	100
Chapter 9 – The Birth of the Royal Air Force	112
Chapter 10 – Summary	122
Bibliography	128

FOREWORD

Foreword

As a third-generation RAF pilot, with a pedigree that stretches in unbroken service from 1915 to 2002, I was invited to set up the WW1 Aviation Heritage Trust on a chance meeting during a visit to the Vintage Aviator Ltd (TVAL) in New Zealand in 2012. The intent was to ensure that the wonderful TVAL aircraft could then be made available in Britain, primarily during the centenary of the Great War, 2014–18. Although scaling back to part-time employment, I was not looking for more work. However, since its inception in 2013, the Trust has become a labour of love and homage. Our pilots feel honoured to fly such faithful reproductions, describing them as 'majestic' (BE2), 'tricky' (Snipe), and 'powerful' (Albatros D.Va). The aeroplanes all have their own quirks, ranging from single magnetos, bungee suspension, no brakes, varying sensitivity to crosswinds and adverse yaw, to name but a few. They must earn their keep, so getting them to an airshow with the right weather conditions for display flying is a lottery. Hidden behind each event are challenges in getting the aeroplanes to the start line – annual servicing, routine and unplanned repairs, Trust governance cycle and the ever-present spectre of them having to complete a journey on a soft-sided truck. However, as I am often told 'if it was easy, everyone would be doing it'. What I had not anticipated was the eclectic mix of amazing characters that I would meet along the way. The reward has been to see the BE2 whispering along at a stately 55mph and the roar of the BR.2a rotary engine as the Snipe went through its paces. The reward can also be seen in the eyes of airshow audiences as they realise how much we owe the aircrews of the Great War fighting in such fragile craft, and how much the foundations of air power were laid in four-and-a-half long years in the cauldron that was northern France more than a century ago.

Wing Commander Dick Forsythe OBE RAF (ret'd)
Chief Trustee of the WW1 Aviation Heritage Trust
2 March 2021

Opposite: Royal Aircraft Factory BE2c operated by the WW1 Aviation Heritage Trust.

PREFACE

Preface

World War One took place just after the birth of powered flight when aerial technology was still in its infancy. At the outbreak of war, the military potential of this novelty invention was not fully recognised. However, some forward-thinking Entente (Allied) commanders soon realised the advantages of using the aeroplane to see what the enemy was planning on the other side of the hill. Once this became clear, the need for up-to-date aerial reconnaissance quickly grew and so too did the need for reliable aircraft.

Commanders from the Central Powers (Germany, Austria-Hungary, the Ottoman Empire and Bulgaria) also sent aircraft over the Entente trenches in search of their own intel, and the need to protect the skies from snooping aircraft became apparent. The role of the aeroplane evolved from a stable, lumbering observation platform to a more agile fighter, capable of shooting down another aircraft. The ongoing arms race in a bid to out-do the enemy saw rapid developments in aeroplane technology. An incredible array of ever-advancing aircraft saw service during this time.

This book features a potted history of the British aircraft that were involved in World War One. It includes the frontline fighters, bombers and reconnaissance aircraft that contributed to the iconic events between 1914 and 1918. The story is supported by high-quality images of surviving reproduction, replica and restored aircraft. Many of the images that feature in this book include airframes with genuine wartime experience.

Acknowledgements

The author and publisher would like to thank the following organisations for permission to use copyright material in this book: The Great War Society, WW1 Aviation Heritage Trust, the Shuttleworth Collection, the Avro Heritage Museum, the Army Flying Museum, Solent Sky Museum, Stow Maries Great War Aerodrome, the Imperial War Museum and the RAF Museums for permissions to photograph their exhibits on their sites. Every attempt has been made to seek permissions for copyright material and photographic rights in this book. However, if we have inadvertently used materials without permission, we apologise, and we will make the necessary correction at the first opportunity.

The author would also like to thank the team at Key Publishing; Georgia Massey for encouragement, motivation and support; Andy 'Loopy' Forester and Daniel Gooch for joining him on aircraft photography adventures. He would also like to acknowledge the support of his close family, especially his father who was with the author when many of these images were taken but sadly passed away in 2020. Finally, the author would like to acknowledge the hard work of airshow organisers, warbird operators, restorers, home builders, conservators, re-enactors and museum curators that keep the memories alive.

Opposite: **The Shuttleworth Trust's Bristol F2b, Sopwith Camel and SE5a in formation.**

CHAPTER 1
INTRODUCTION

Introduction

At the start of the World War One, the Royal Flying Corp took just four squadrons of primitive aircraft to France. In 1914, their only purpose was to gain information. Pilots soon found that they would have to fight for this information, and so began the first crude duels in the air. As the fighting intensified, a technological war began to run parallel to the one above the trenches. The need for ever more sophisticated aircraft grew as each side developed their own purpose-built fighters known as 'scouts'. The potential military purposes of ever-improving aircraft designs soon became apparent to both sides, and by 1918 there were different flying machines specifically built for bombing, day-fighting, night-fighting, escorting and reconnaissance.

These early machines were not built to last; they were mostly built from extra lightweight materials to compensate for low engine power. As aeroplane technology developed so quickly, there seemed little point in keeping any of these outdated fragile wood and fabric structures. Although some aircraft were adapted for civilian purposes, most were scrapped just after the war. There was also little thought for preservation at that time, which makes original surviving World War One aircraft very rare indeed. Fortunately, some aircraft have survived and been restored to their original conditions for display in our museums. We are also privileged to have an ever-growing replica and reproduction scene, which allows us to see what some of these aircraft would have been like in the air.

Opposite: A line-up of a Sopwith Triplane, Sopwith Camel and Bristol M1c prepare to take to the skies at Old Warden.

Right: The Shuttleworth Collection's crew return the Sopwith Camel to the hangar following a display. Meanwhile, the SE5a flies overhead.

A lone pilot was vulnerable to attack, so they soon began to work in pairs. Pairs became squadrons, and, eventually, wings of up to 50 machines would work together to protect each other as they crossed into enemy territory. These large wings could be made up of single-seater scouts, two-seater-fighting machines and the all-important photoreconnaissance aeroplane. The need to outclass the enemy in the air drove the aircraft industry to rapidly improve its technologies. At the start of the war, power flight was little over ten years old and aircraft manufacturers were still debating and testing the best designs. Often, aircraft were rushed into service with very deadly consequences only to be withdrawn shortly afterwards. The range of aircraft that saw service during the conflict is staggering.

This image shows some of the experimentations at the time: a Bristol M1c and Sopwith Triplane reproduction take to the skies together at Old Warden. Although the monoplane would dominate the future of aviation at the time of World War One, the single wing often proved flimsy or was unable to provide sufficient lift for the low-powered engines of the time. The Sopwith Triplane proved successful for a brief period after its introduction into the war. Its three wings provided considerable lift, manoeuvrability and a rigid structure but also more drag.

Debates on how to best lay out an aircraft raged on during the early stages of the war. The position of the propeller in relation to the engine was one of the most hotly contested topics. Many aircraft, including the Wright Flyer that took the first flight in 1903, used a 'pusher' configuration where the propeller was mounted behind the engine to push the aircraft through the air. Other aircraft, including the BE2 (pictured above left), used a 'tractor' layout where the propeller was at the front of the aircraft pulling it through the air.

The Vickers Gunbus (pictured above right) benefited from a pusher layout, as it meant that the pilots and observers had an unobstructed view and, most importantly, machine guns could be fired forwards. It quickly became apparent that the tractor layout had more potential, and from around 1916 the development of a mechanism for synchronising a machine gun with propeller blades conceded the main advantage of pusher aircraft. Although experimentation with pusher aircraft would continue well into World War Two, the tractor layout quickly became the predominate design.

envisioned and the debate on where best to fit the gun continued well into the war. In fact, some very early designs lacked the power to even carry the weight of the gun. The arrival of the propeller-synchronised machine gun in the middle of the war largely settled this debate.

The Great War Display Team (GWDT) is a group of experienced pilots who display replica World War One aircraft on the UK airshow

As the war progressed, it became apparent that air-to-air combat was going to be critical in ensuring that reconnaissance could successfully be gathered. An arms race to find the most suitable platform for aerial combat soon began. After a few early pilots took unsuccessful pop-shots at their flying rivals with hand pistols, machine guns were fitted to aircraft as standard. Most aircraft were designed before this was

circuit. They perform daring low-level dogfight routines with smoke and pyrotechnics that aim to replicate how these aircraft were flown over the trenches during the war. Here we see the Royal Aircraft Factory BE2 taking a hit from the chasing Fokker Dr I. A Royal Aircraft Factory SE5a can also been seen in the background.

The layout of World War One aircraft differed significantly; as wing designs and engine positions were hotly debated, so too were the number of crew. A single-seat aircraft could be faster, lighter and more agile, but two seats provided better protection and enabled the pilot to concentrate on flying the aircraft whilst the observer could take photographs, look out for the enemy or, if need be, fire the guns.

The two images here allow comparison between the two-seat Bristol F2b and the single-seat Royal Aircraft Factory SE5a. Both aircraft proved successful in the later stages of the war. Despite differences in design and gun layout, they were both used in similar roles. These two airworthy examples belong to the Shuttleworth Trust and are regular performers at the Old Warden airshows.

A line-up of three World War One aircraft at Old Warden: Avro 504, Sopwith Triplane and Sopwith Pup.

Following four years of war, the world's population was keen to move on as soon as peace was finally declared on 11 November 1918. Demobilisation was fast; men and women were soon thrown back into their civilian lives and military technology was tossed swiftly onto the scrapheap. The war was fought at great costs to both sides; the losses could be measured both economically and in the tragic, unprecedented loss of human lives. It was no surprise that little thought was given to preservation of machinery at this time.

Incredibly, some genuine aircraft from this period have survived. Although source figures vary, it is thought that around 330 original wartime aircraft are now preserved across the world. Today, having recently experienced the 100-year anniversary of the Great War, we are also fortunate to be experiencing a renewed interest in World War One aviation, and there are now many more replica and reproduction aircraft on show for us to see.

An Avro 504 displays with a Bristol F2b.

World War One saw the birth of the flying ace. The likes of Billy Bishop, Edward 'Mick' Mannock and James McCudden did battle against enemy pilots such as Manfred von Richthofen, Oswald Boelcke and Max Immelmann. Their goal was to score the highest number of aerial victories and become the ace of aces. In the process, they became celebrities and heroes, giving a sense of glamour and chivalry to the air war. The realities of aerial warfare at the time were very different; the average life expectancy of a new pilot was fewer than three weeks.

In the first year of the war, almost half of the fatalities of British aircrew were attributed to training mishaps or mechanical failures rather than actual combat. Even without the war, flying during the early days of aviation was a hazardous business. Although the aircrew lived well behind the front line in relative comfort compared with those in the trenches, it is hard to comprehend the danger they were in when taking to the air. Aircraft technology was in its infancy and it would have to grow up fast.

CHAPTER 2
EARLY AVIATION IN BRITAIN: PRE-WAR AIRCRAFT

Early Aviation in Britain: Pre-war Aircraft

Balloons, airships and kites were the first craft to carry people above the battleground to gain a better view of the opposition. Hot-air balloons were utilised for military observation from as early as the American Civil War in 1862. The British military also experimented with aerial observation and, in 1906, the man-lifting kite was adopted by the Royal Engineers. They were generally tethered and could raise an observer to a height of up to 1,500ft (457m). They were in service until 1912 when heavier-than-air-powered machines began to creep into service offering more flexibility.

This replica of a man-lifting kite can be viewed at the Army Flying Museum in Middle Wallop. It represents just one segment of the kite, which would have been built up of nine of these segments to lift the observer and his basket into the air. These were designed and produced by the Royal Balloon Factory at Farnborough; this would later become the Royal Aircraft Factory. This book will focus on powered aeroplanes, although it should be acknowledged that lighter-than-air aircraft such as balloons and airships were widely used by both sides during World War One.

Opposite: The Shuttleworth Collection operates several 'Edwardian' aircraft including the Bristol Boxkite and Avro Triplane replicas built for the *Those Magnificent Men in their Flying Machines* movie.

After the Wright brothers took the first powered flight on 17 December 1903, other countries soon developed their own aircraft. Initially, Britain was lagging in development; in Europe, the French took the lead. Louis Blériot designed a series of aircraft that proved successful in the early days of aviation. He was the first to fly across the English Channel in 1909 using his self-designed Blériot XI. Although it was a fragile design that barely made the trip over the Channel, it was still in military service for France, Britain, Russia and Italy at the beginning of the war.

The Humber bicycle and motor car factory in Coventry obtained a licence to build up to 40 of the French-designed aircraft in Britain. The first Humber Blériot XIs appeared in 1910 and were powered by a

30hp three-cylinder Humber engine based on the original Anzani design. This replica of the Humber Blériot (pictured left) can be seen at the Midland Air Museum in Coventry; the Humber engine is an original. The Humber company would continue to produce aero engines on behalf of Bentley for the Air Ministry throughout the war. There are several other French Blériots on display in the UK, including some modern replicas and originals at the RAF Museum (pictured above) and Shuttleworth Collection. The latter is thought to be the oldest airworthy aircraft in the world.

Whilst other European nations embraced powered flight, it took a little while for the British aircraft industry to find its feet. In 1910, the British and Colonial Aeroplane Company in Bristol started to experiment with designs. Its first successful aircraft was based on a French imported Zodiac biplane that was designed by Henri Farman. The machine was a very close copy but just different enough to avoid court proceedings. The aircraft soon became known as the Bristol Boxkite because of its appearance; it equipped the new flying schools at Brooklands and Larkhill. The Royal Naval Air Service (RNAS) adopted the Boxkite in 1911; the Army Air Battalion would also take some into service the following year. Although cumbersome to fly, they would continue into service until 1915 when they would be replaced with machines more suited to warfare.

This replica Boxkite is owned and operated by the Shuttleworth Trust in Old Warden, Bedfordshire. It was built by the Miles Aircraft company in the 1960s for use in the *Those Magnificent Men in their Flying Machines* movie. This example is powered by a Rolls-Royce continental engine with 100hp as opposed to the 50hp Gnome engine that would have been fitted to the originals. The aircraft is still flown today when conditions are calm enough for a safe flight.

The first powered flight in Britain was undertaken in 1907 by American pilot Samuel Cody in an aircraft based on the Wright brothers' design known as the British Army Aeroplane No. 1. Two years later, Alliott Verdon-Roe became the first British subject to design and fly an all-British aircraft. The fight took place at Lea Marshes, Walthamstow, on 13 July 1909. The aircraft was the Roe Triplane I, powered by a 9hp JAP motorbike engine. The project was funded by his brother Humphrey, and it made several successful flights that year. A remarkable achievement, considering that Roe had never received any instruction on how to fly an aeroplane. Trading as A V Roe or Avro, Alliott Verdon-Roe's newly formed company would go on to great success designing hugely successful and iconic aircraft for the British military for many years to come.

The original Triplane is on display in London's Science Museum, but this replica was built by the staff at the Avro Heritage Museum in Woodford. It was sponsored by Eric Verdon-Roe, the grandson of Alliott Verdon-Roe. The aircraft was built for flight and even boasts a genuine JAP engine. Sadly, the goal of flying the replica was not achieved, as it was not possible to modify the plans sufficiently to meet Civil Aviation Authority guidelines. It can still be viewed at the museum and will have its wings fitted at some point in the future.

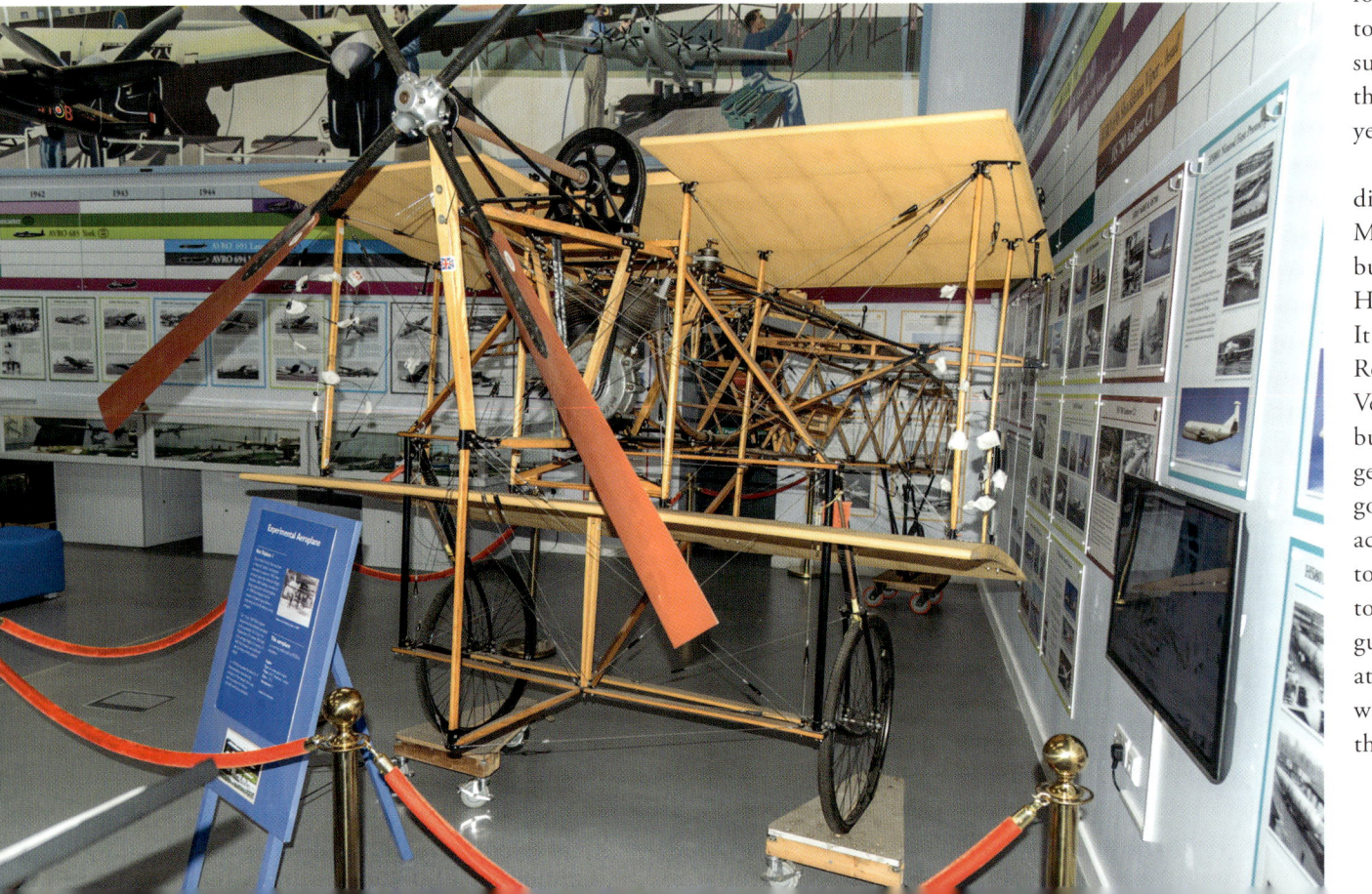

Alliott Verdon-Roe would go on to form the famous Avro company. Initially, he continued to upgrade his triplane design and would eventually produce four versions of this layout. At the time, the triplane idea was unusual; it enabled a greater wing area, producing more lift with a lower-powered engine and thus keeping the weight down as much as possible. The machine featured lateral control through wing warping via a steering wheel on top of a control column in the cockpit. The column also worked the elevators, making it one of the first aircraft with a single control column.

Although this aircraft design was not put into mass production, it gave the Avro company vital experience in aircraft design that would be used in its future successful aircraft designs such as the Avro 504. The aircraft pictured here is a replica Avro Triplane IV built for the *Those Magnificent Men in their Flying Machines* movie. The replica is considered authentic to the original but is powered by a Cirrus-Hermes engine instead of the Green C.4 that would have been on the original. It can still be seen flying at the Shuttleworth Collection airshows when weather conditions allow.

Another British aviation pioneer Robert Blackburn was also tinkering with aircraft designs before World War One. In 1909, he established his own workshop in Leeds. He favoured the monoplane design initially and, like Verdon-Roe, he began making a series of ever-improving aircraft. His designs were known as the 'Mercury' series and by 1912 he had produced a successful variant that was attracting a small number of customers. Although this aircraft would not see mass production, it provided a foundation for the Blackburn company to go on producing aircraft for the British military for over 50 years.

The aircraft pictured here is an original; it was built in 1912 for a private owner. After being put into storage during World War One, it was forgotten until 1937 when it was uncovered and eventually sold to Richard Shuttleworth who would begin its lengthy restoration to flight. Sadly, Richard Shuttleworth would not live to see it return to the skies, but the trust formed in his honour now flies this aircraft on a regular basis. It is the oldest airworthy British aircraft anywhere in the world. It is seen here flying in all its glory and in a stripped back state during a recent restoration and overhaul.

CHAPTER 3
EYES IN THE SKY: RECONNAISSANCE AIRCRAFT

Two re-enactors from the Great War Society recreate a flight planning scene at Stow Maries Great War Aerodrome in Essex.

When World War One began, Britain sent just four Royal Flying Corp (RFC) squadrons of aircraft to France. These included an eclectic mix of aircraft types including Blériots, Avro 504s, FE2s, BE8s and BE2s. Most were two-seater aircraft and they were all intended solely for gathering reconnaissance. With this intention in mind, the BE2 was considered the most capable aircraft on charge at the time. It was stable in flight, cruised at around 3,000ft, and with a speed of approximately 60mph it was not too fast – all excellent characteristics for observing and photographing the lie of the land, providing those being observed had no objection.

Over 3,000 BE2s were produced, of which only three originals have survived: one in Canada, one in France and this one in the UK. This BE2c (2699) can be viewed at the Imperial War Museum, Duxford and appears in a clear dope finish. It served with the No. 50 Home Defence Squadron at Dover and flew several night sorties before being transferred to No. 190 Night Training Squadron in April 1918. It completed its service life with No. 51 Squadron before being withdrawn from service at the end of 1918. The aircraft has enjoyed many years on display for the Imperial War Museum at both the Duxford and Lambeth sites.

Between 1911 and 1918, the Royal Aircraft Factory (formerly the Royal Balloon Factory), based at Farnborough produced several different aircraft for military purposes. It was the establishment's official aircraft research department; several of the aircraft designed at Farnborough entered mass production during the war. The Royal Aircraft Factory was responsible for supplying aircraft to the Army's own air service, the RFC. Meanwhile, the RNAS took its pick of the privately produced aircraft

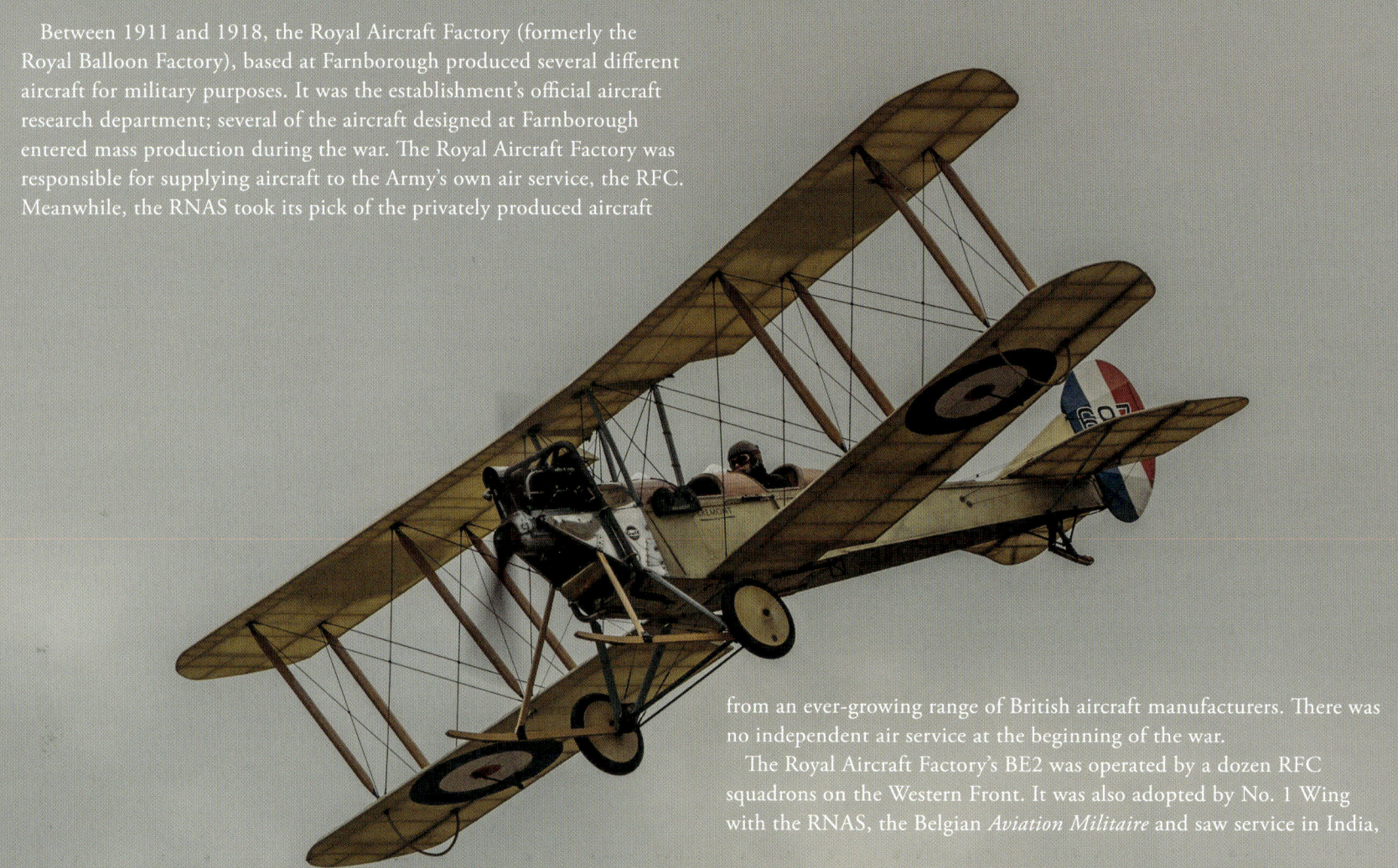

from an ever-growing range of British aircraft manufacturers. There was no independent air service at the beginning of the war.

The Royal Aircraft Factory's BE2 was operated by a dozen RFC squadrons on the Western Front. It was also adopted by No. 1 Wing with the RNAS, the Belgian *Aviation Militaire* and saw service in India,

Africa, Australia and the Aegean. This flying replica BE2c was built in 1969 by David and Charles Boddington for the *Biggles Sweeps the Skies* movie that was never actually made. The aircraft was flown, crashed and stored for many years in the USA before being brought back to the UK and returned to flight by Matthew (son of Charles) Boddington and Steve Slater. It has been a popular airshow attraction as part of the GWDT since 2011 but recently suffered a crash in September 2020. Matthew Boddington was flying the aircraft at the time but was largely unscathed; it is unknown if the aircraft will fly again at this point.

The Royal Aircraft Factory developed its own system for classifying aircraft: each new design was classified with initials based on its similarities to pre-war foreign aircraft types. For example, the BE aircraft were named Blériot Experimental as their 'tractor' layout with the engine at the front of the aircraft was the same as the original Channel-hopping Blériot. Pusher aircraft, with the engine behind the pilot, were designated Farman Experimental (FE) after the French Farman brothers' early designs. There was also an SE designation for Santos Experimental

Above and opposite: The Great War Society re-enact scenes from the period at Stow Maries Great War Aerodrome with a replica BE2c

after Santos Dumont's canard layout with the tail at the front. However, these types were not widely used and later in the war SE became short for Scout Experimental.

Although not well suited to the front line, the BE2 proved successful as a home defence aircraft, and the stability proved an asset for night flying. For home defence, some BE2s were converted to single-seaters and given a forward-firing Lewis gun (with this configuration, five German airships were destroyed). This flying BE2e replica (A2943) is in No. 7 Squadron of the RFC colours. The squadron used BE2s for reconnaissance and bombing during the Battle of the Somme in 1916. The original A2943 is recorded as being flown by Capt Horace Webb-Bowen. It is owned and operated by the WW1 Aviation Heritage Trust based at the remarkably intact Great War Aerodrome in Stow Maries, Essex.

Some of the airframes developed by the Royal Aircraft Factory went on to achieve great success; others barely got off the drawing board. The BE2 was the outcome of Edward Busk's experiments on keeping aircraft stable in flight; in 1914 this was deemed a desirable quality for collecting reconnaissance. The BE2c was put into mass production and initially proved to be very good at its job. However, Germany and the Central Powers soon began to develop better fighters, and the BE2 crews had little hope of defending themselves or evading the opposition. Sadly, it took a long time to replace the BE2 on the front line, and over 60 BE2 aircrew lost their lives during 'Bloody April' in 1917 alone.

This replica BE2b was built using some original metal fittings taken from the remains of at least three original aircraft. The rest of the airframe was built following original factory drawings in the late 1970s and early 1980s; the engine is an exact copy of the one on display at the Science Museum. It wears the colours of BE2b 687, which was flown by 2nd Lt William Barnard Rhodes-Moorhouse on 26 April 1915. He set off on a bombing run on Courtrai railway station in northern France but was badly wounded by rifle fire and splinters from his own bomb. He managed to return home to his base airfield at Merville but died the following day. He was awarded a posthumous Victoria Cross. This aircraft is now on display in the Graham White hangar at the RAF Museum, Hendon.

On 22 August 1914, pilot 2nd Lt Vincent Waterfall and his navigator, Lt Charles George Gordon Bayly, took off in an Avro 504 on a reconnaissance flight over Flanders, Belgium. Tragically, this aircraft became the first British aeroplane to be shot down by the Central Powers in the war. Both crew members lost their lives in the incident. Following this and other similar incidents, the Avro 504 was deemed unsuitable for frontline duties but found its niche in other roles, becoming one of the most successful training aircraft to ever be produced.

The first Avro 504k (pictured above left) is a reproduction that was constructed from original manufacturer's drawings by Hawker Restoration. It was produced for a private client but is currently operated by the Imperial War Museum at Duxford. The second Avro 504k (pictured above right) is the amalgamation of two original airframes: the fuselage was taken from Avro 504k G-EBJE and the wings are from Avro 548A G-EBKN. The Avro 548A was an aircraft developed later by Avro, but heavily based on the 504. This aircraft is based at the RAF Museum in Hendon.

The Avro 504 was the first biplane produced by the now famous Avro company. It was the most produced aircraft of World War One, and by the time production had ceased in 1932, over 10,000 had been built. It took its first flight at Brooklands in 1913 in the hands of Fred Raynham and, like the BE2, was designed for stable reconnaissance flights. Both the RFC and RNAS operated the Avro 504 at the beginning of the war and took small numbers of the machine to France as soon as war was declared.

The 504 was one of the first aircraft that the then young designer Roy Chadwick would work on during his time with Avro. Chadwick would later go on to design some of the most iconic British aircraft of all time, including the Avro Lancaster. This airworthy replica Avro 504k (G-EROE) was built in Argentina and is now owned by Eric Alliott Verdon-Roe (grandson of Alliott Verdon-Roe). G-EROE is currently represented by the Navy Wings charity, which pays homage to British naval aviation. It has appeared at several airshows in recent years.

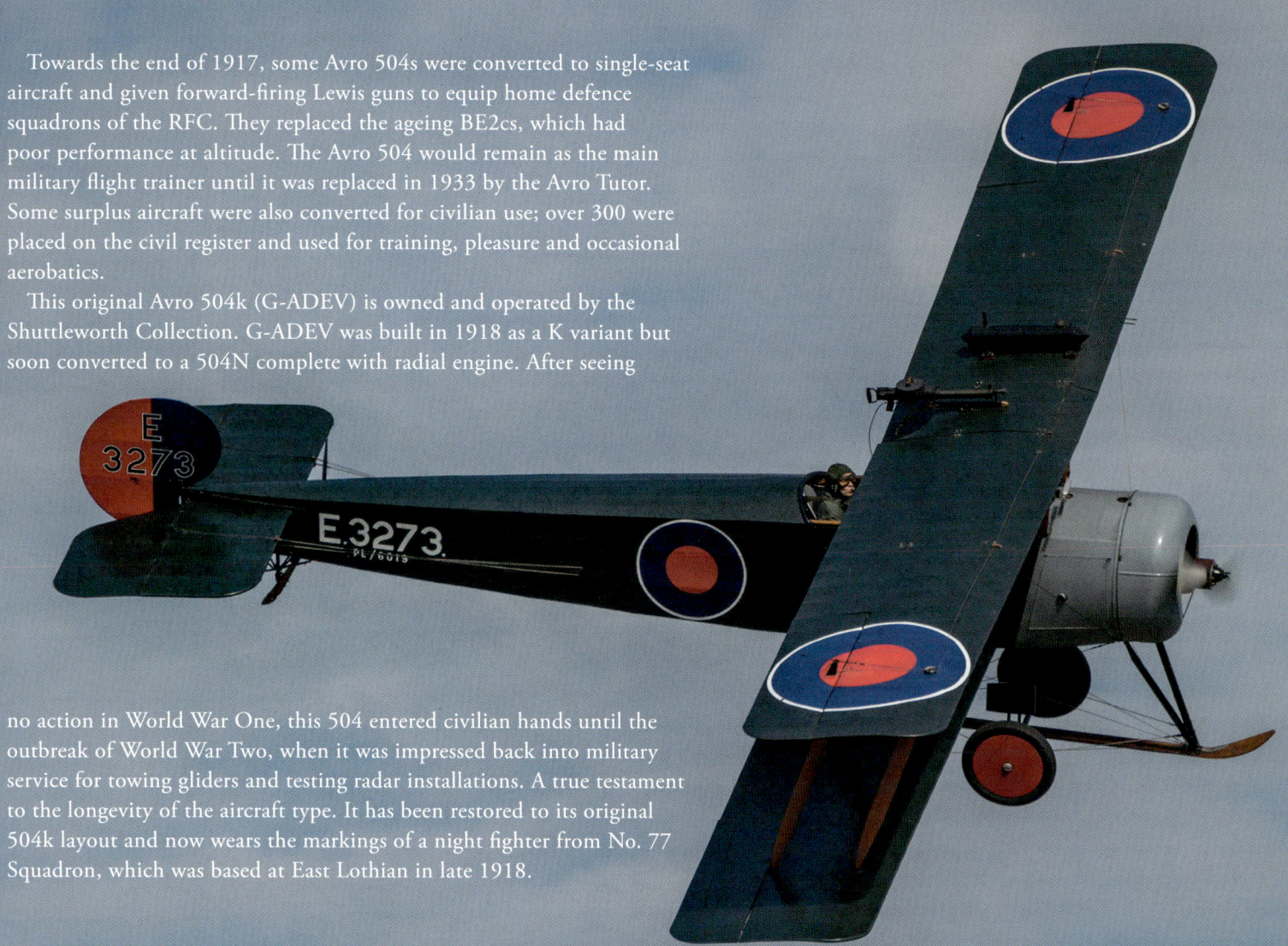

Towards the end of 1917, some Avro 504s were converted to single-seat aircraft and given forward-firing Lewis guns to equip home defence squadrons of the RFC. They replaced the ageing BE2cs, which had poor performance at altitude. The Avro 504 would remain as the main military flight trainer until it was replaced in 1933 by the Avro Tutor. Some surplus aircraft were also converted for civilian use; over 300 were placed on the civil register and used for training, pleasure and occasional aerobatics.

This original Avro 504k (G-ADEV) is owned and operated by the Shuttleworth Collection. G-ADEV was built in 1918 as a K variant but soon converted to a 504N complete with radial engine. After seeing no action in World War One, this 504 entered civilian hands until the outbreak of World War Two, when it was impressed back into military service for towing gliders and testing radar installations. A true testament to the longevity of the aircraft type. It has been restored to its original 504k layout and now wears the markings of a night fighter from No. 77 Squadron, which was based at East Lothian in late 1918.

The Royal Aircraft Factory designed the RE8 (Reconnaissance Experimental) as a replacement for the struggling BE2. Over 3,000 were built, of which the majority saw action over the Western Front for the RFC. Unfortunately, it offered little improvement over the BE2. It maintained an inherent stability, which was good for artillery spotting but was offset by a lack of manoeuvrability in a dogfight. The 'Harry Tate', as it became known, was also susceptible to engine failures, fires and spinning at low altitude. Over time, improvements were made to both the aircraft design and the way its pilots were trained. The RE8 provided a valuable reconnaissance service throughout the war.

Two original RE8s have survived the tests of time: one is on display at the Brussels Air Museum in Belgium, whilst the other, F3556 (pictured here), is currently hanging from the ceiling in the Airspace hangar at Duxford. It features the inscription 'A Paddy Bird from Ceylon' on the front fuselage as the newspaper *The Times of Ceylon* spearheaded a sponsorship scheme to help pay for aircraft. The name was passed on from a BE2 that was lost in 1916. F3556 did not arrive in France until Armistice Day and, as such, saw no action in the war.

The RE8 was powered by a Royal Aircraft Factory 4a air-cooled V12 engine. It was capable of just over 100mph and could operate at up to 13,000ft. On 21 November 1916, No. 52 Squadron of the RFC became the first squadron to take the RE8 into battle. The squadron was operational with the RE8 over France, but after several accidents it soon switched back to the BE2e. The first decisive action involving RE8s was during the Battle of Messines in June 1917 where the aircraft were able to successfully guide counter-battery fire helping to destroy 72 enemy batteries. By October 1918, the RE8 was the most widely used British reconnaissance aircraft and was even adopted by the officer commanding the RFC in the field, Brig Gen Hugh Trenchard, as a personal transport.

TVAL in New Zealand has built four reproduction RE8s, including one that can be seen in the UK (A3930). It was built in 2011, and was shipped to the UK to be flown by the Shuttleworth Collection. After a brief period of flight, it was semi-retired and put on static display at the RAF Museum in Hendon. The original A3930 was flown by No. 9 Squadron of the RFC on the Western Front from May 1917.

CHAPTER 4
THE BIRTH OF THE SCOUT: EARLY SINGLE-SEATERS

This replica Sopwith Pup makes an interesting backdrop for the Great War Society re-enactors recreating an early war aviation scene at Stow Maries Great War Aerodrome in Essex.

The Birth of the Scout: Early Single-Seaters

The very first reconnaissance aeroplanes were generally unarmed. However, it was not uncommon for pilots to carry pistols or rifles for use in self-defence or even hunting if they were downed and left stranded, particularly on the Eastern Front. Aircraft of the period were also prone to fire, and in the absence of a parachute, the pistol provided pilots and aircrew with a quick way out should they choose to take it.

As the skies over the front grew busier, it was not long before aircrew would use their light weapons to take pop-shots at their rivals as they flew past. Some airmen even carried grenades with the intention of dropping them onto the enemy below. However, using these light weapons proved largely ineffective. It was clear from the opening stages of the war that more effective, purpose-built fighting aircraft would be required.

Initially based on weight, single-seat aircraft were deemed the most suitable for taking on enemy aircraft in the air. They were faster and more agile than the standard two-seat reconnaissance platforms in service in 1914. These new single-seat fighting aircraft were known as 'scouts', as it would be their role to seek out enemy aircraft to prevent them from gathering their own intelligence. Scouts, such as the Sopwith Pups, would also be deployed to escort larger reconnaissance aircraft and to ensure that the Entente controlled the skies.

Re-enactor Jed Jaggard demonstrates early air-to-air combat in this replica Sopwith Pup.

The war in the air became increasingly aggressive. Aircrew soon realised they would have to fight for their right to gather aerial intelligence. Even after they had run out of ammunition in their handheld weapons, pilots would dive on and circle their opponents to intimidate them.

On 25 August 1914, the RFC achieved its first aerial victory in this way when three BE2 aircraft from No. 2 Squadron forced down a German Taube. Although this was not air-to-air combat as we know it today, it was a stark reminder that aeroplanes needed to be armed and protected.

As the first few months of the war progressed, crude air-to-air combat slowly became more sophisticated. The first British single-seat Scout aircraft to appear on the front line was the Bristol Scout. It was designed by aeronautical engineer Frank Barnwell using well-thought-out design principles, many of which still hold true today. Produced by the British and Colonial Aeroplane Company (universally known as Bristol), it was capable of top speeds just shy of 100mph and was armed with two rifles set on either side of the cockpit that were aligned to fire off at 45 degrees to miss the propeller arc. It proved quite a challenge for pilots to find their mark.

Bristol Scout 1264. Note the Union flag, which was soon replaced by the roundel to avoid misidentification with the German Iron Cross, which would appear similar at a distance.

Aerial combat had been foreseen before the war, but owing to the low power of early aero engines, the weight of an aircraft was at a premium. The machine gun was the obvious weapon of choice, but many aircraft at the time simply lacked sufficient power to get off the ground with one fitted.

The first British aircraft fitted with machine guns were low-powered pusher aircraft that could take over 30 minutes to reach a fighting altitude; clearly ineffective if an interception was required.

Capt Lanoe Hawker of No. 6 Squadron of the RFC experimented with replacing the rifles with Lewis machine guns on his Bristol Scout. Although the challenging angle of fire remained, Hawker downed two German aircraft on 25 July 1915 using this configuration. He was awarded the Victoria Cross for his actions over Passchendaele and Zillebeke. He was the first heavier-than-air pilot to receive such an accolade. The Bristol Scout remained an effective frontline Scout for the first 18 months of the war but would soon be outdated by the next generation of German aircraft.

Bristol Scout 1264 was built by brothers David and Rick Bremner and their friend Theo Willford. It was built as a tribute to David's and Rick's grandfather, Flt S Lt F D H 'Bunny' Bremner who flew Bristol Scout 1264 with the RNAS No. 2 Wing in the Eastern Mediterranean during World War One. This aircraft (1264) is the only airworthy Bristol Scout C in the world. It is a reproduction aircraft built to the original designs and incorporates three original parts that were found in Bunny Bremner's garage after his death in the early part of 1983.

It seems that Bunny was so proud of his aircraft that he took the control stick, the rudder bar and magneto as souvenirs after the war. These are the only original Bristol Scout parts known to exist. The rest of the aircraft was built following the original blueprints and using authentic materials and techniques wherever possible. Bristol Scout 1264 is powered by an original 90hp Le Rhône rotary engine like the one that would have been used in the original. The aircraft has also received a seal of approval from Sir George White whose signature appears on the propeller. He is the grandson of his namesake who founded what became known as the Bristol Aeroplane Company in 1910. The aircraft is now based at Old Warden in Bedfordshire and has appeared at several airshows and commemorative events since its first flight in 2016.

It quickly became clear that aiming a gun in one direction and an aeroplane in another was an ineffective means of aerial combat. Pusher aircraft such as the Airco DH2 benefited from a forward-firing machine gun, but the engine layout prevented this type of aircraft from achieving the speed and manoeuvrability required for combat. Tractor aircraft such as the Sopwith Pup (pictured) were much more suitable for air-to-air

dogfights, but angling a machine gun to fire around the propeller arc while within easy reach of the pilot proved problematic.

This Sopwith Pup replica was restored using a £5,000 grant from Southend-on-Sea Borough Council and is based at Stow Maries Great War Aerodrome. The aerodrome, near Maldon in Essex, was established in 1916 as the base of the No. 37 (Home Defence) Squadron of the RFC. It is the only complete surviving World War One aerodrome in the UK. This non-flyable replica wears the colours of Capt Claude Ridley DSO MC, who was the first commanding officer of the squadron.

Pilots became weary with trying to manoeuvre their aircraft into awkward positions to line up the angle of fire with enemy aircraft. The need for a machine gun that could fire directly forwards was obvious. French pilot Roland Garros took the first initiative, working with Raymond Saulnier of the Morane-Saulnier aeroplane company who was working on a synchroniser to enable the machine gun to fire through the propeller arc. Although this was not quite flawless and refined, Garros insisted on testing it in combat. He devised metal bullet deflectors to fit onto his propeller blades to protect his aircraft from wayward bullets. The test proved a success, but sadly Garros was downed just a few weeks later and his aircraft was captured for analysis.

The machine was eventually passed on to Dutch aeroplane designer Anthony Fokker, who was already working on his own device. The result was a successful cam-operated synchroniser that the German authorities

The Birth of the Scout: Early Single-Seaters

insisted Fokker test himself in the field. Fokker found himself unable to fire the trigger and the testing was left to less-scrupulous German airmen. The device was fitted to Fokker Eindecker monoplanes and, in the hands of German aces like Max Immelmann and Oswald Boelcke, proved a formidable foe. The Allies soon recognised the peril they were in and so began the 'Fokker Scourge', which would continue until the summer of 1916 when aircraft such as the Sopwith Pup began to arrive with their own synchronised machine guns.

This Sopwith Pup was built in 1919 and converted to a two-seater civilian aircraft known as the Sopwith Dove. The aircraft eventually found itself in the hands of Richard Shuttleworth, who set about returning the aircraft to its original Pup configuration. The aircraft remains airworthy today and currently wears the markings of Pup 9917 that served with the RNAS on HMS *Manxman*.

The Fokker Eindecker that caused such devastation during the first half of 1916 was not considered a remarkable aeroplane. In fact, the Bristol Scout was considered more than its equal but lacked the all-important synchronised machine gun. When the first Sopwith Pups reached the Western Front in sufficient numbers, the balance was redressed. This small, manoeuvrable aircraft allowed the Allies to dominate the skies until the middle of 1917. One of the greatest British aces James McCudden believed that the Pup could turn twice as fast as the Albatros D.III (a later variant is pictured opposite), its closet German rival.

Four original Sopwith Pups survive today. This one, N5195, is on display at the Army Flying Museum at Middle Wallop. Along with its RAF Museum cousin, it was restored by Lt Cdr Desmond St Cyrien in the 1960s. It was eventually made airworthy in 1985 but now enjoys semi-retirement as a static exhibit. The Albatros pictured opposite is a D.Va replica built by TVAL in New Zealand, but it has recently returned to the UK as part of the WW1 Aviation Heritage Trust's collection.

The RNAS enjoyed considerable success with the Pup it equipped the Seaplane Defence Flight with, which protected shipping and escorted slower seaplanes on reconnaissance work. It was also used for trials in deck flying by the Navy. Although taking off from an aircraft carrier proved straightforward at that time, there was no means to land back on the deck. Instead, the Pups were fitted with flotation devices that helped them to be hoisted back on board after ditching. The RNAS employed several other floatplanes throughout the war, including the Short 184 and the Sopwith Baby; this was a viable alternative to carrier operations whilst landing back on deck remained problematic.

The Sopwith Pup enjoyed a short but productive career. Its Le Rhône 9C 80hp engine made it capable of over 100mph. It had a range of over 300 miles and it could remain airborne for up to three hours. The Pup was popular with its pilots and was used for many duties on the front line, home defence and later as a successful trainer.

Sopwith Pup N6161 was built in 1916 and served in France until a forced landing on 1 February 1917. The aircraft was then captured intact and flown by the Germans for analysis of its flying characteristics. The aircraft has recently been restored to flight by Retrotec, which managed to utilise many of its original parts combined with other original Sopwith items. After taking its first post-restoration flight in the UK on 17 October 2016, it has now been shipped to its permanent residence in America.

The Sopwith Pup first flew in February 1916 and was immediately adopted by the RNAS as the Sopwith Type 9901. After seeing its performance reports, the RFC also adopted the aircraft, which it officially called the Sopwith Scout. However, when it reached the front line, the aircrews noticed its lineage to the larger Sopwith 1½ Strutter and nicknamed it the 'Pup'. Despite decrees from above from those who thought the name Pup undignified, the name stuck and has become universally accepted.

This original Sopwith Pup (N5182) (pictured on the left) sits next to a replica Sopwith 1½ Strutter – the resemblance is clear. N5182 was built in 1916 and initially taken up by the RFC but quickly transferred to the RNAS where it carried out most of its service. It flew several bombing escort patrols and fighting patrols throughout the last two years of the war. On 23 November 1916, N5182, flown by Australian ace Robert Alexander Little, scored a victory against an unidentified enemy aircraft. Little would score a further two victories in this aircraft in December that year and would finish the war as Australia's highest-scoring ace.

Biplanes were very popular during World War One; the extra wing provided more lift, which aided the low-powered engines of the time. The struts and additional wing also gave the aircraft added stability; the quest to ensure aircraft were as light as possible often compromised rigidity and could cause aircraft to break up in flight. This led to a bias in the ministry that prevented the wider uptake of monoplanes, even if their performance was equal to that of the biplanes at the time.

The Bristol M1c was designed as a monoplane Scout. It was completed in 1916 and powered by a 110hp Clerget rotary engine that gave it a maximum speed of 130mph, considerably faster than most of its contemporaries. There is only one surviving original aircraft, which is on display in Minlaton, Australia. There are, however, two excellent replicas in the UK. This one is now on display at the RAF Museum in Cosford, and was built as an airworthy replica in 1987 and given a slightly more modern Warner Scarab radial engine.

The Air Ministry ordered 130 Bristol M1cs, which began service in the RFC towards the end of 1916. However, scepticism over monoplanes meant that this aircraft type was not used on the Western Front. It was largely reserved for training, but, eventually, five squadrons were equipped for active duties in the Middle East. Despite its restricted service, the aircraft seems to have performed well; Capt Frederick Dudley Travers DFC of No. 150 Squadron achieved ace status in this aircraft but is the only pilot to have done so.

This reproduction Bristol M1c was built by volunteers at the Northern Aeroplane Workshop. It is a reproduction built following original plans and using authentic techniques and materials where practical for a modern flying aircraft. It was completed in 1997 and delivered to the Shuttleworth Collection where it remains in immaculate, airworthy condition. After the team at Old Warden completed a few finishing details and checks, it made its first flight on 25 September 2000.

The Bristol M1c was not widely used during World War One, but when the Royal Air Force was formed, it would become the first monoplane fighter in service and would point the way forward for the future.

CHAPTER 5

AN EXTRA PAIR OF EYES: TWO-SEATER AIRCRAFT

The Shuttleworth Collection's Bristol F2b.

An Extra Pair of Eyes: Two-Seater Aircraft

At the beginning of World War One, the two-seater aircraft was deemed the only suitable layout for a reconnaissance aircraft. The pilot could concentrate on flying, whilst the observer could gather the intelligence by making notes, sketches or, later, photographs. As the air war turned deadly, the observer became a gunner, in an aircraft layout reminiscent to a naval battleship. An aircraft could carry a rear-gun turret, which enabled the aircrew to defend themselves from several angles.

Early two-seater aircraft such as the Royal Aircraft Factory BE2 (pictured below left) proved vulnerable to attack from the front or below. Communication was also an issue between the aircrew, as the gunner did not know where the pilot was going to turn next, making targeting enemy aircraft problematic. The introduction of more advanced, faster, agile two-seaters such as the Bristol F2b (pictured below right) fitted with forward-firing machine guns proved successful later in the war. Providing the aircraft's performance could match a single-seater, having an extra pair of eyes and more guns was always an advantage.

An Extra Pair of Eyes: Two-Seater Aircraft

Designated as the Sopwith Type 9700 by the RNAS or the Sopwith Two-Seater with the RFC, this aircraft was better known by its nickname, the Sopwith 1½ Strutter. It had an unusual arrangement of central main plane bracing struts, which earnt it its popular name. It was the first British aircraft to be fitted with a synchronised machine gun. When first introduced to the Western Front in early 1916, it briefly had the edge over German fighters until the Albatros and Halberstadt began to appear just a few months later.

Around 1,500 Sopwith 1½ Strutters were produced in Britain, with a further 4,500 being produced in France. They were utilised in all theatres of World War One. Some were even converted to single-seaters, but the smaller, more agile Pup exceeded their performance in this domain. Only four original 1½ Strutters survive. None of these are on display in the UK, but the RAF Museum hosts this former airworthy reproduction at its Cosford site. It was built from original factory drawings in 1980 and currently wears the markings of A8226, which was in service for the RFC during 1917.

The Vickers FB5 (Fighting Biplane 5) was the first British aircraft purpose-built for air-to-air combat. The quest for an aircraft capable of destroying other aeroplanes in the air was begun by Vickers in 1912. It produced a handful of fighting biplane designs before arriving at the FB5, which became known as the Gunbus. The Gunbus made use of the pusher layout, which meant that a gunner could be positioned right at the front of the aircraft with an uninterrupted view and field of fire. It performed reliably for most of 1915 but eventually proved too slow to be an effective interceptor.

Around 200 FB5s were produced, but no originals survive today. This aircraft (G-ATVP) was built as an airworthy replica in 1966 by the Vintage Aircraft and Flying Association at Brooklands. It was painted in the markings of an FB5 2345 (Bombay 2) and is currently displayed at the Hendon site of the RAF Museum. The original 2345 was flown by Capt Hereward de Havilland in November 1915. His observer, 2nd Lt C K M Douglas, was an amateur meteorologist. Together, the crew became the first to collect meteorological data in an aeroplane. Douglas fitted a thermometer to the gun mounting and used the data to research cloud formations.

The Bristol F2 Fighter was developed in 1915 to replace the Royal Aircraft Factory's BE2 two-seater reconnaissance aeroplane that was suffering terrible losses to German Scout pilots during the opening stages of the war. Also known as the 'Brisfit', the F2 took its first flight in September 1916 and after overcoming some early teething problems, it proved to be a quantum leap forward in performance over the BE2.

Eight original Bristol F2s have survived; three of these remain airworthy including one based in the UK with the others in Canada and New Zealand. This example (E2581) is considered to be the most authentic surviving Bristol Fighter. It was displayed at the Imperial War Museum's Lambeth site for many years, before being moved to Duxford in the 1980s. E2581 was built by the British Colonial Aeroplane (Bristol) Company in September 1918. It was issued to No. 39 Home Defence Squadron at North Weald in Essex right at the end of the war. It served for the newly formed Royal Air Force until it was transferred to the museum in 1923.

The definitive Bristol Fighter was the F2b variant, which was fitted with a Rolls-Royce Falcon III engine. It was capable of 123mph and could stay airborne for over three hours – an impressive performance for a large two-seater aircraft. The F2b was so effective that even as a two-seater it could match the German single-seater scouts in a dogfight. By the end of the war, over 5,000 F2s had been built and many of these were converted for civilian use during peacetime.

This F2b is on display at the RAF Museum in Hendon. It is a well-crafted reproduction aircraft that utilises several original components salvaged from around six different airframes and even boasts a reproduction Rolls-Royce Falcon engine. The starboard side is left uncovered, offering a unique insight into its construction. The aircraft has been given the markings of No. 22 Squadron aircraft, E2466, which was flown in France by Capt W F John Harvey and his observer, Capt D E F Waight, in the summer of 1918.

entered service in 1917, it was operated cautiously by its pilots who were used to lumbering two-seaters. It was not until they learned to fly it more like a fighter that its fearsome reputation was earned.

This Bristol F2b (D8096) is based at Old Warden airfield and was built in 1918. It was a little too late to serve in the war, but it did serve for the RAF in Turkey in 1923 with No. 208 Squadron. After some time in storage, the Bristol Aeroplane Company restored the aircraft to flight in 1952. It has been with the Shuttleworth Trust ever since. The Bristol F2b proved to be one of the most successful aircraft of the war and continued in military service for many years afterwards.

The Bristol Fighter was designed from the ground up as a fighter-reconnaissance aeroplane. Although it is a large aircraft compared to other contemporary fighters, its Rolls-Royce III V12 engine provides 275hp, giving it an exceptional performance for the time. When it first

An Extra Pair of Eyes: Two-Seater Aircraft

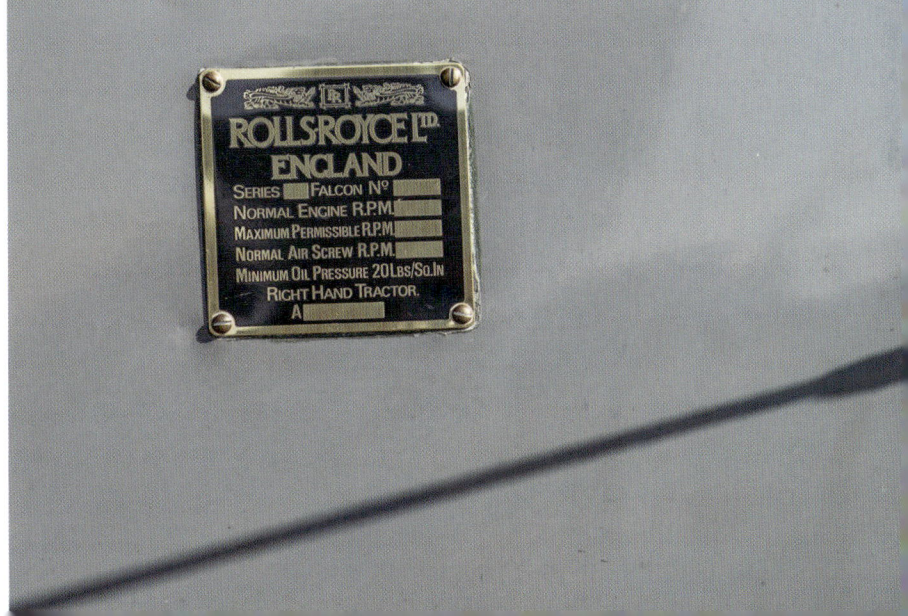

CHAPTER 6
THE AIR WAR MATURES: ADVANCED SINGLE-SEATERS

The Shuttleworth Collection's reproduction Sopwith Camel.

The Air War Matures: Advanced Single-Seaters

Despite new Entente fighters arriving on the front line, the Germans maintained air superiority for the first few months of 1917. The Germans had devised new tactics and the superb Albatros D.III fighter continued to wreak havoc in the hands of vastly experienced pilots including Baron von Richthofen. The RFC's response was to send aircraft over in greater numbers than ever; a single reconnaissance aeroplane would be given five escorts. However, it was all in vain, as 245 British aircraft were lost in battle, killing 211 aircrew in just one month. That month became known as Bloody April.

An increasing amount of pressure was put upon British aircraft designers to come up with a fighter capable of besting the enemy. Many different configurations were tested during a period where aviation was still in its infancy. Success had been achieved early on with biplanes or monoplanes, but in early 1917 the Sopwith company developed the successful Pup fighter into a triplane. When introduced onto the front line, it immediately got the better of the Albatros. The Germans soon followed with their own triplanes, but despite further experiments including the unsuccessful Wight Quadruplane, British aircraft manufacturers soon returned to more conventional biplane layouts.

The GWDT's Sopwith Triplane and Fokker Dr I.

Solent Sky Museum's replica of the ill-fated Wight Quadruplane that crashed on its first test flight.

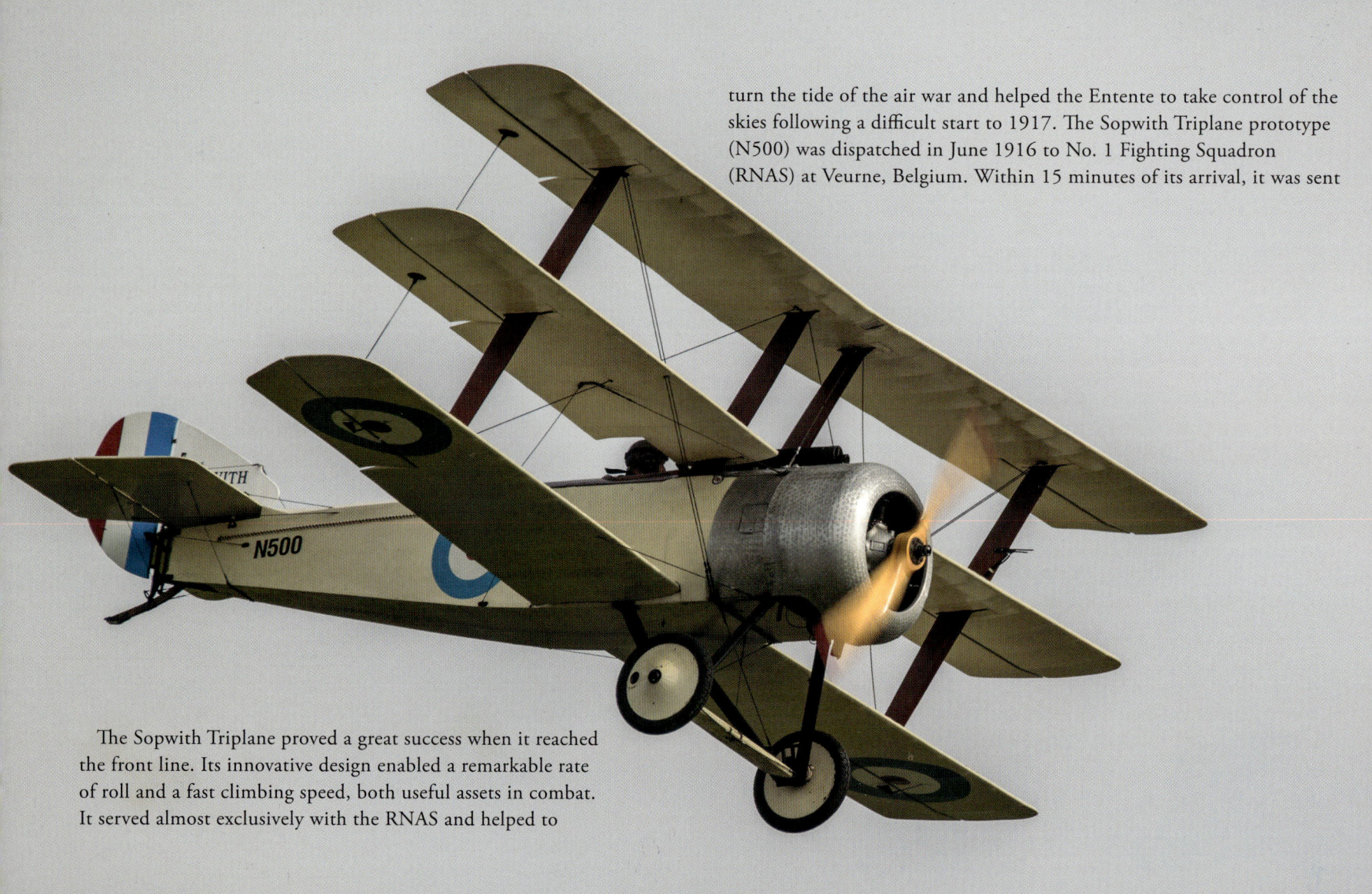

turn the tide of the air war and helped the Entente to take control of the skies following a difficult start to 1917. The Sopwith Triplane prototype (N500) was dispatched in June 1916 to No. 1 Fighting Squadron (RNAS) at Veurne, Belgium. Within 15 minutes of its arrival, it was sent

The Sopwith Triplane proved a great success when it reached the front line. Its innovative design enabled a remarkable rate of roll and a fast climbing speed, both useful assets in combat. It served almost exclusively with the RNAS and helped to

up in pursuit of an enemy aircraft – not the usual way to test an aircraft but a clear demonstration of the desperate situation at the time.

The GWDT, which is currently sponsored by Bremont, flies dynamic displays at low level with pyrotechnics and an eclectic mix of Great War replica aircraft. In its ranks it boasts a replica Sopwith Triplane that is

currently painted to represent N500. It appears in the clear dope finish that it would have worn on its first visit to the Western Front. The aircraft is currently owned by Gordon Brander who has been displaying it for over ten years. The team also displays two Fokker Dr I replica triplanes, seen in the picture here with the Sopwith Triplane and BE2c. One of the Dr Is is owned by Iron Maiden front-man Bruce Dickinson. The Fokker Dr I arrived much later on in the war and never actually met the Sopwith Triplane in battle.

The Sopwith Triplane was quickly put into production and entered service with No. 1 and No. 8 RNAS Squadrons in February 1917. No. 10 Squadron would also be equipped two months later. It retained the single gun and general layout of the Sopwith Pup but was given a more powerful 130hp Clerget engine and an extra wing. Between May and July 1917, triplanes from No. 10 Squadron were responsible for destroying 87 enemy aircraft. It was not surprising to see Austrian and German designers copying the idea and putting their own triplanes into production. Despite a fleeting moment of success, the triplane layout created additional drag, and aircraft designers soon returned to biplanes.

The Shuttleworth Collection owns and operates this reproduction triplane that was delivered to Old Warden in 1990 and made its first flight two years later. It was built to original plans by the Northern Aeroplane Workshop and includes an original 130hp Clerget engine. Sir Thomas Sopwith (founder of the Sopwith and later Hawker Aviation companies) inspected the aircraft and declared it a 'late production' aircraft rather than a replica. It now represents N6290/Dixie II of No. 8 RNAS Squadron.

The Triplane was developed by Herbert Smith at Sopwith's experimental department. It was a private venture but soon gained backing by the RNAS. The Sopwith company was contracted to build 95 triplanes, but Clayton and Shuttleworth also produced around 40 aircraft as a sub-contractor. Despite early success and high praise, the aircraft was ultimately flawed: it proved hard to maintain in the field and was subject to structural failure. The major drawback, however, was its limited armament; the incoming Camels and contemporary German aircraft were boasting two machine guns.

Only two original Sopwith Triplanes are thought to survive. One is on display in Russia, whilst this one, N5912, in currently on display at the RAF Museum in Hendon. It was built in 1917 but did not see any active service. Instead, it was used as a training aircraft by both the RFC and RNAS until 1919. After passing through the hands of the Imperial War Museum, Science Museum and Fleet Air Arm Museum, it eventually ended up at the newly created RAF Museum in 1971, and it has been there ever since.

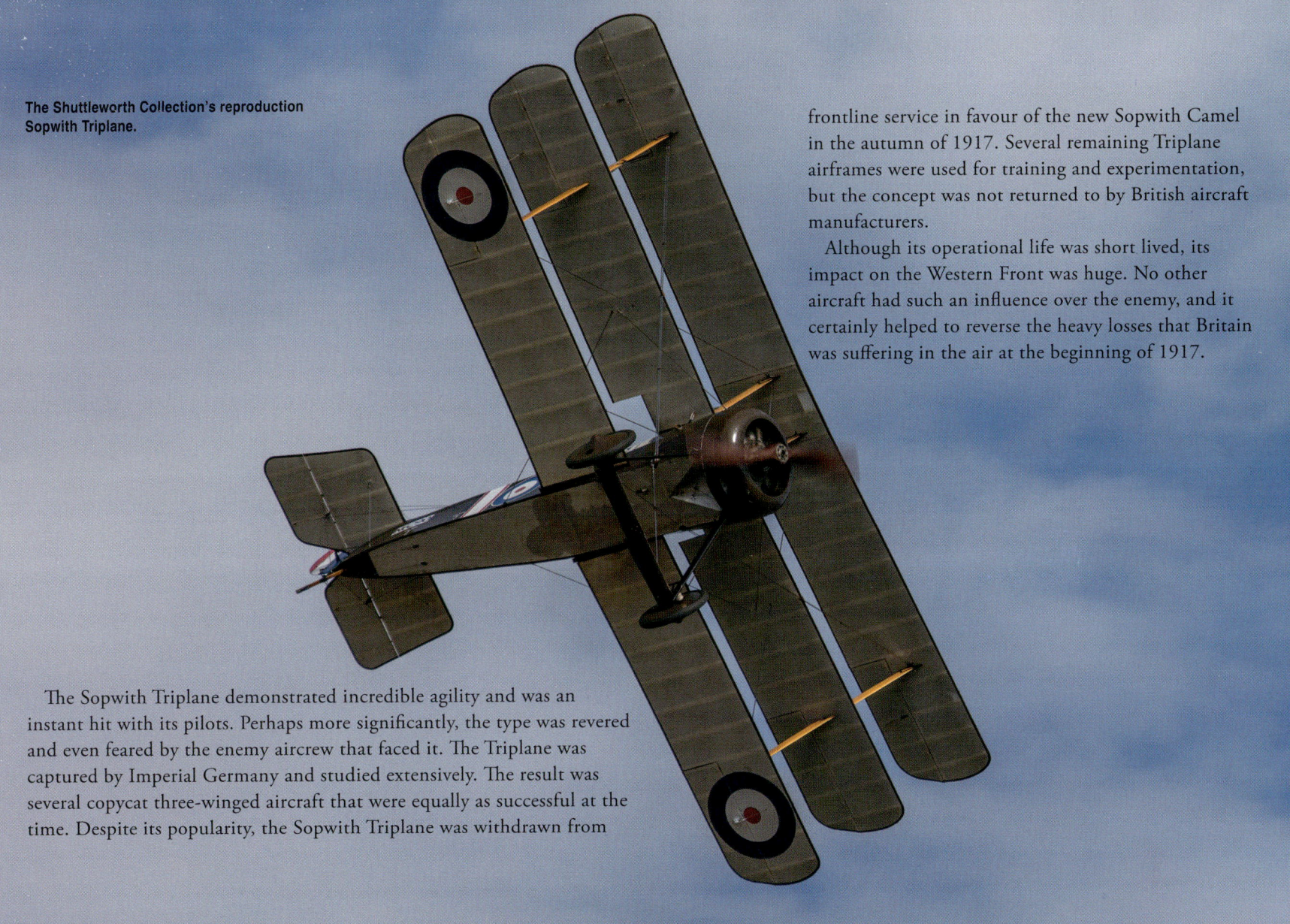

The Shuttleworth Collection's reproduction Sopwith Triplane.

The Sopwith Triplane demonstrated incredible agility and was an instant hit with its pilots. Perhaps more significantly, the type was revered and even feared by the enemy aircrew that faced it. The Triplane was captured by Imperial Germany and studied extensively. The result was several copycat three-winged aircraft that were equally as successful at the time. Despite its popularity, the Sopwith Triplane was withdrawn from frontline service in favour of the new Sopwith Camel in the autumn of 1917. Several remaining Triplane airframes were used for training and experimentation, but the concept was not returned to by British aircraft manufacturers.

Although its operational life was short lived, its impact on the Western Front was huge. No other aircraft had such an influence over the enemy, and it certainly helped to reverse the heavy losses that Britain was suffering in the air at the beginning of 1917.

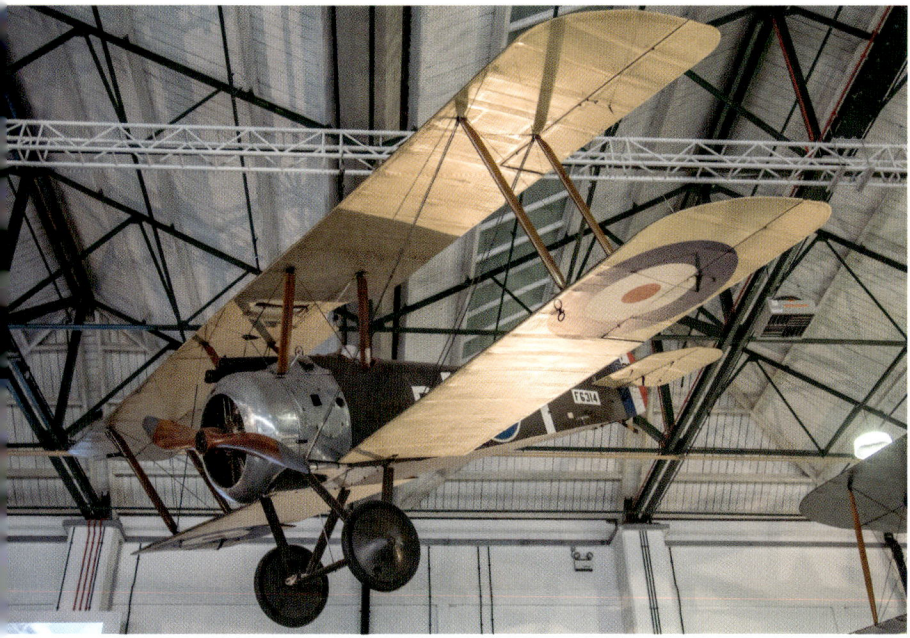

The Sopwith F1 was the next step in the evolution of Sopwith's incredible line of World War One Scout aircraft. It was Britain's first production fighter to feature twin synchronised machine guns – the cowling over which gave the aircraft a distinctive hump giving rise to its now universally accepted nickname, the Camel. It has been likened to the Spitfires of World War Two as *the* iconic aircraft of the period. It was responsible for bringing down almost 1,300 enemy aircraft, including three airships – the largest tally for any single aircraft type during the war. The Camel was employed in many roles during the war: it was an excellent air-to-air fighter; it was also employed in home defence squadrons and could even be equipped with four Cooper bombs under the wing and employed as what was known as a 'trench fighter' in a ground-attack role.

The RAF Museum houses an original Sopwith Camel (F6314), which was built by Boulton & Paul Ltd in 1918. It was not used during the war but was rescued from the Aircraft Disposal Company in 1923 by former Camel pilot Grenville Manton. After a largely unsuccessful attempt to return the aircraft to flight, Manton put the aircraft up for sale. It passed through several owners, but F6314 eventually found its way into the Nash collection at London Airport where it received a restoration and an update in markings to represent No. 65 Squadron of the RFC. It toured the country as part of several RAF displays in the 1960s but has been at the RAF Museum in Hendon since its opening in 1972.

The Air War Matures: Advanced Single-Seaters

The Sopwith Camel first took to the skies towards the end of 1916. Initially, it was powered by a 130hp Clerget rotary engine, but other power plants were used as production expanded. It was capable of 115mph and could operate at up to 21,000ft. Its powerful right-hand torque radial engine enabled the aircraft to make instant snap rolls to starboard. Once pilots had learned to exploit this, the Camel was unrivalled in its agility. The Camel first entered service with the RNAS in June 1917 before being taken on by RFC one month later. Well over 5,000 Camels were produced and they eventually made up 50 RFC and 8 RNAS squadrons. The Camel helped the Allies regain air superiority over the Western Front.

There are thought to be eight original Camels in existence today; two of these are currently on display in the UK, including the one at the RAF Museum (on the opposite page) and another at the Imperial War Museum in London. In addition to the original aircraft, there is an ever-increasing population of replica and reproduction Camels across the world. The Camel pictured here is an airworthy reproduction owned by the Shuttleworth Collection.

Although the Sopwith Camel proved a formidable aircraft in the hands of an experienced pilot, it could be tricky to handle for those unprepared for its startling performance. Its incredible manoeuvrability was such a rapid departure from its docile predecessors that it proved just as deadly for the hastily trained, fresh-faced airman as it did to the enemy. The engine torque and forward centre of gravity meant that the aircraft tended to climb when rolled left, and to descend when rolled right. It needed constant left rudder input to counteract the engine torque to maintain level flight. Therefore, inexperienced pilots were often caught out, resulting in the deaths of 413 Camel pilots in combat and a further 385 losses in non-combat accidents.

This reproduction Sopwith Camel was built by Northern Aeroplane Workshops in Batley, West Yorkshire, for the Shuttleworth Collection. Work started in 1995 and was eventually completed when the aircraft took its first flight in May 2017. The aircraft was made using traditional methods and materials and even includes original wheels, fuel tanks and a restored 140hp Clerget engine. It is painted to represent D1851 IKANOPIT (I can hop it), which was part of No. 70 Squadron of the RFC and, later, the RAF in 1918. It is a regular flyer at the Shuttleworth airshows and often teams up with some of the other World War One aircraft based there.

The Air War Matures: Advanced Single-Seaters

The Sopwith Camel (pictured opposite) earned its place in history when, in 1918, Canadian pilots Wilfred May and Roy Brown encountered the famous German ace Manfred von Richthofen, known as the 'Red Baron' because his aircraft were painted red all over. The Red Baron had spotted Wilfred May and singled the newly qualified pilot as an easy target. However, as Richthofen set off in pursuit of May, Brown joined the chase. He shot several short bursts of fire towards the German ace who made a forced landing and succumbed to his injuries shortly afterwards. The Australian infantry were also shooting up from the ground and it is unclear if they or the Canadian pilot Roy Brown were responsible for the final shot.

This replica Sopwith Camel was built by Viv Bellamy at Westward Airways in 1977. It was initially airworthy and registered as G-BFCZ. However, it has not flown since 2003. Its markings represent B7270 of No. 209 Squadron of the RAF, the aircraft flown by Capt Roy Brown when he downed the Red Baron. It is now based at Brooklands Museum where it often runs up its engines for visitors.

This replica Dreidecker Fokker Dr I is painted in the markings worn by the Red Baron. It is pictured at the Imperial War Museum in Duxford. The engine visible in the second picture was taken directly from Richthofen's aircraft after he was shot down on 21 April 1918. It is a Oberusel UR2, which was a German copy of a French design. Richthofen scored 80 victories during World War One; the last 19 were in the famous red triplane represented here.

CHAPTER 7
THE MOUNT OF ACES: ROYAL AIRCRAFT FACTORY SE5A

The SE5a entered service during the later stages of the Great War. It was developed by the Royal Aircraft Factory at Farnborough and was one of the fastest aircraft of the time. Although the Sopwith Camel is possibly the best-known fighter of World War One, the SE5a was flown by almost all the highest-scoring British aces, including Maj Edward C 'Mick' Mannock, who scored 50 of his 75 kills in it. The SE5a had excellent speed and impressive range and was well-armed with its two .303 machine guns. It was highly manoeuvrable and, unlike the Sopwith Camel, was safe even for inexperienced pilots.

The GWDT has operated several ⅞ths scale replica SE5as, which have all been built following plans devised in the 1960s by the Canadian Replica Plans Company. The first example to fly in Britain was G-BDWJ, which currently flies in the colours worn by No. 85 Squadron of the RFC. More specifically, it represents F8010, which is thought to have been flown by Maj Mannock. It was built in 1978 by Mike Beach and is now flown at airshows by former RAF Harrier pilot Dave Linney.

Opposite: The Shuttleworth Collection's SE5a – the only airworthy original World War One aircraft to have scored an aerial victory during the war.

Above: G-BDWJ, a ⅞ths scale replica SE5a flown as part of the GWDT.

The SE5a had a top speed of 138mph and a range of 300 miles, considerably better than most of its closest allies and rivals at the time. Despite its success on the front line, its development was far from straightforward. The first 'SE' machines were designed by Geoffrey de Havilland who suffered a serious accident whilst test flying the SE2. The crash left de Havilland with a broken jaw, and it effectively ended his time with the Royal Aircraft Factory. Henry Folland then picked up the development of the series, which eventually culminated in the SE5.

The Shuttleworth Collection is proud to home an impressive array of significant aircraft, but this SE5a (F904) is truly special. It is the only aircraft that achieved an air-to-air victory during the war that is still airworthy. On 10 November 1918, Maj C E M Pickthorn MC successfully destroyed a Fokker D.VII just over Chimay in Belgium. This was his last of five victories, giving him ace status on the final full day of war.

Following the war, F904 was purchased with several other SE5s (three of which still survive today) by Maj J C Savage for his skywriting

business. Following a decade of use, the aircraft was put into storage until its eventual restoration that now includes a refurbished 200hp Wolseley Viper engine. It is currently displayed in the markings of No. 84 Squadron of the RAF. Maj Pickthorn was in command of this squadron for the last few days of the war. Although F904 flies regularly for the Shuttleworth Collection's airshows, it is rarely seen away from Old Warden. Understandably, the Shuttleworth Trust is keen to handle this aircraft with extreme care.

Replica SE5a, C1096, currently based at Old Warden airfield.

The first SE5 was powered by a 150hp Hispano-Suiza V8 monobloc engine. In this guise, the aircraft was heavily criticised by famous ace Albert Ball. Ball made a few valid points when it came to the SE5's shortcomings, even though he was probably swayed by his father's interests in a rival machine being developed by Austin Motors. It is also possible that Ball's faith in the Royal Aircraft Factory's designs had been shaken by his earlier experiences in the BE2. He had so far favoured the French Nieuport aircraft and achieved many of his kills in this machine. One of the suggestions adopted onto the SE5 included the fitting of a Lewis gun on the section of the upper wing as seen in the Nieuport designs.

A replica Nieuport 17 now owned by the WW1 Aviation Heritage Trust collection.

Albert Ball would go on to achieve 12 of his 44 confirmed kills in the SE5. Tragically, it was also the aircraft he was killed in. The details of Ball's death are uncertain, but he was last seen on 7 May 1917 pursuing a German Albatros thought to have been piloted by the Red Baron's younger brother, Lothar von Richthofen. The Germans eventually dropped a message over the front line announcing that Ball was dead and had been buried in Annœullin with full military honours.

Most of Ball's other kills were achieved in a Nieuport 17, a French fighter design with a sesquiplane layout, the lower wing much smaller than the top one. The similarities in design and layout between the Nieuport 17 and SE5a can be seen in the images here. Thanks to Ball, both shared a similar machine-gun layout, as can be seen here on these topside views.

The markings of this replica Nieuport 17 are dedicated to N1977 of the Lafayette Escadrille, flown by Sgt Robert Soubiran. The Lafayette Escadrille was a French squadron flown by American volunteers. It is now powered by a Warner Scarab radial engine and is owned and flown by John Gilbert as part of the WW1 Aviation Heritage Trust collection.

The early production woes continued when Farnborough test pilot Frank Goodden was killed in an SE5 after a catastrophic wing failure at 1,500ft. Following the suggestions of Albert Ball and the results of the investigation into Goodden's fatal accident, the SE5 was modified with several changes. These changes included a more powerful 200hp Hispano-Suiza engine and a shorter wingspan. Sadly, these modifications would be too slow coming for some. No. 56 Squadron of the RFC was the first to take the SE5 into battle, and the aircraft was rushed into service in April 1917 to help rebalance the air war over the Western Front. Although losses were heavy, the experiences of the pilots were fed back, and the aircraft was soon modified into the world-beating SE5a.

SE5a, F938 (formerly G-EBIC), is currently displayed at the RAF Museum in Hendon. It was built by Wolseley Motors in Birmingham in 1918. It is the second of three surviving SE5as that were purchased by Maj Savage after the war. The other one now resides at the Science Museum in London. Like many of the exhibits now on display at Hendon, F938 became part of the Nash collection before the RAF Museum was built. It now wears the colours of No. 56 Squadron, the first unit to take the SE5 into battle.

The Mount of Aces: Royal Aircraft Factory SE5a

Although the SE5a was considered by many as the best aircraft of the time, it was initially unable to have the same impact as the more famous Sopwith Camel and Pup owing to a shortage of Hispano-Suiza engines. The engine would be later produced under licence in the UK as the Wolseley Viper, which helped to bring the SE5a to the front line in much greater numbers. Once it was widely available in the last few months of the war, it is thought that it helped to bring a quicker end to the conflict.

James McCudden was the third highest-scoring British ace of the war. He achieved a total of 57 victories, of which 49 were achieved flying the SE5. Like Albert Ball, he had an excellent understanding of how his aircraft worked and made several modifications to his SE5 to enable better performance at altitude. Most of his modifications were made personally and unrecorded, but it is thought that he was able to climb to 10,000ft five minutes faster than any other SE5. Sadly, McCudden was killed in an accident on a return flight to France in July 1918. He had taken a much-needed break after suffering combat fatigue. McCudden had promised his fiancée that he would not do anything foolish on his return to action and yet it is felt that had he worn his safety harness, he may have survived the crash.

This replica SE5a (G-BFWD) is based around a Phoenix Currie Wot biplane and is owned by David Silsbury and Brendan Proctor.

Another SE5a replica representing No. 56 Squadron has recently emerged after a lengthy restoration process. It was first built in the late 1980s by John Tetley in his workshop in Yorkshire. In the early part of the 21st century, it was transferred to the French Memorial Flight in a deal that allowed them to fly it for ten years in exchange for completing the build with their own Hispano-Suiza engine. In 2014, it returned to the UK and was modified to replicate the aircraft flown by Canadian ace Henry John 'Hank' Burden during the war. Burden achieved 16 victories during the war, all whilst flying the SE5a from 1918 onwards. On 10 August 1918, he achieved five victories in one day, shooting down five Fokker D.VIIs; he would claim another three two days later.

The aircraft flown in his honour is as authentic as possible in every respect. It incorporates many original parts including instruments, fuel tank, Lewis gun and magazine. The undercarriage was rebuilt using Hank Burden's notes, and the markings now accurately represent the No. 56 Squadron aircraft flown by Hank in 1918. The aircraft is now based at the Shuttleworth Collection and is likely to fly alongside its original SE5a at future events.

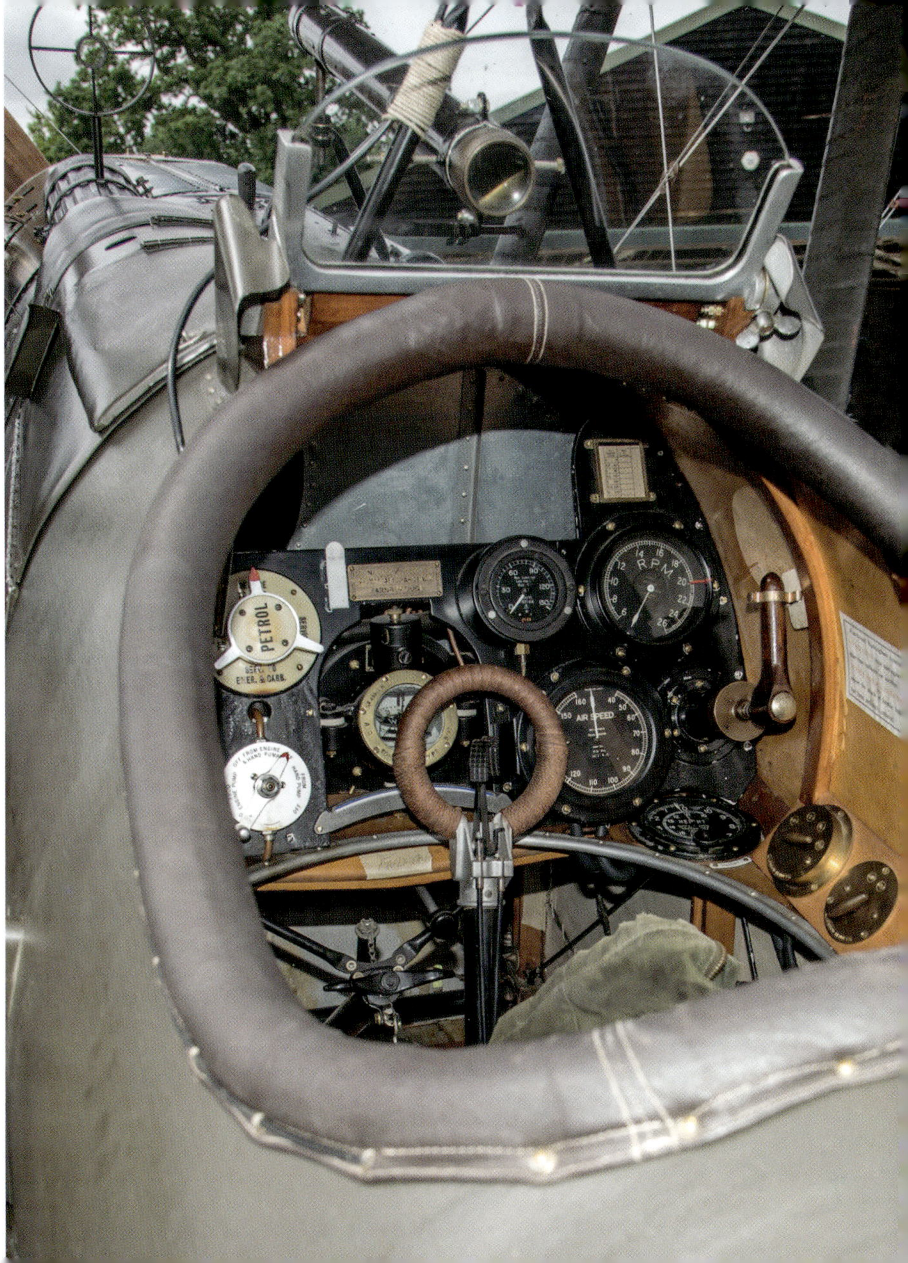

By the summer of 1917, the Germans had begun probing bombing raids into England using Gotha bomber aircraft. The first raid over Folkestone could only be met with lumbering BE2s and BE12s, which proved ineffective. It was not until the German bombers turned their attention to London that the British government finally stepped in and ordered back No. 56 Squadron and their SE5s to shore up the home defences. Poor weather meant that the squadron saw little action over England but it was glad of the month's relief from the front line. After their return to France, Nos 61 and 143 Squadrons took over the home defence duties using the SE5a, which proved effective until the Germans switched to night-time raids.

Stow Maries Great War Aerodrome lies to the east of Chelmsford in Essex. It is the largest known surviving group of RFC buildings that has never been adapted for modern use. It was opened in 1916 as the home of RFC B Flight of No. 37 Squadron that took charge of the air defence over Britain in response to the increasing threat of Zeppelins and Gotha Bombers. This SE5a replica (G-BMBD) was built by Des Biggs but is currently owned by David Blaxland. It is seen here recreating history with the Great War Society re-enactors at Stow Maries Great War Aerodrome.

SE5a replicas, from left to right: G-BUOD, G-BDWJ, G-CCBN and Matthew Boddington's BE2c.

One of the GWDT's SE5a replicas evades two Fokker Dr Is.

One of the GWDT's SE5a replicas narrowly avoids the anti-aircraft fire nicknamed 'Archie' at the time.

William 'Billy' Avery Bishop was the top scoring Entente fighter ace of World War One. He was credited with 72 victories and survived the war. The Canadian initially trained as an observer and excelled in aerial photography, soon becoming responsible for training others. After a brief period as an observer, Bishop learned to fly and began his aerial fighting career in a BE2c. After flying several types on the front, he was eventually given the SE5a where he continued to notch up his kill tally. Despite a lengthy period away from the front line in a number of significant training roles, he would finish the war as the third highest-scoring ace, just behind von Richthofen and René Fonck.

The GWDT is a popular airshow attraction. It often performs mock dogfights and makes great use of pyrotechnics to replicate anti-aircraft fire, known as 'Archie' during World War One. The team operates several ⅞ths scale replica SE5as, three of which are pictured in the static shot opposite with the BE2c in the background. The team formed in the late 1980s and is made up of a collection of civilian-owned aircraft and their enthusiastic owners and pilots. The team aims to give some idea of what aerial conflict was like during World War One.

The GWDT's 7/8ths scale replicas G-BUOD and G-INNY.

The GWDT's ⁷/₈ths scale replica SE5a G-CCBN.

Above: The GWDT's ⁷/₈ths scale replica SE5a G-BDWJ.

Below: The GWDT's ⁷/₈ths scale replica SE5a G-BUOD.

As well as G-BDWJ seen earlier in this chapter, the GWDT's SE5a replicas include G-CCBN, which appears here as 80105 of the US Air Service. The unusual American markings represent Blue 19 of Aero Squadron based at Toul. It is currently owned by Arnd Schweisthal and often flown by Emily or Mike Collett.

G-BUOD can also be seen here displaying with the team in the markings of B595 of No. 56 Squadron. This aircraft represents the one flown by ace Lt Maurice Edmund Mealing MC who achieved many of his 14 victories in an SE5a. This aircraft was built by Mike Waldron in 1995 and is now flown in the team by Trevor Bailey. Finally, G-INNY has also recently returned to the team. It was built by Mark Ordish and is occasionally flown by Mark Johnson.

George McElroy was born in County Dublin, Ireland in 1893. He enlisted as soon as World War One began and after two years' service in the Royal Irish Regiment he was transferred into the RFC. He benefited from a close relationship with Mick Mannock who mentored the young pilot and passed on several tips to aid his aerial combat. McElroy first flew the Nieuport 17 but was unable to achieve a victory until he transferred to the SE5a. He claimed his first victory on 28 December 1917, and would go on to achieve 46 more before his untimely death in combat in July 1918.

This replica SE5a is pictured outside of Ragley Hall in Alcester, Warwickshire. It is the ancestral home of the Marquess of Hertford and was used during World War One as a military hospital, which helped to save the hall from impending demolition at the time. Today, Ragley Hall is a popular attraction and is visited by numerous tourists every year. Most recently it has played host to the Midlands Air Festival where this shot was taken.

The SE5a saw extensive service in all theatres throughout World War One and remained in RAF service for some time after the war. Many SE5as also continued flying in civilian hands and have proved popular with home builders and modern pilots today. It was an aircraft that had few vices and, for the time, was easy to master even for the inexperienced pilot.

The SE5a proved to be one of the most successful aircrafts of the war. It certainly helped the Entente retake control of the skies over the Western Front, and although it did not achieve the same fame as the Sopwith Camel, it was flown by many of the highest-scoring aces of the time. Highly decorated pilots including Edward 'Mick' Mannock, James McCudden, George McElroy and Albert Ball all flew the SE5a and achieved over 40 victories in aerial combat. It is no wonder the SE5a is known as the 'Mount of Aces'.

The Shuttleworth Collection's original SE5a in action over Old Warden.

CHAPTER 8
TAKING THE FIGHT TO THE ENEMY: BOMBING AIRCRAFT

Military leaders soon realised that the aeroplane (or airship) was not just limited to use as a reconnaissance tool. Despite the technological immaturity of aircraft in the early stages of the war, both sides began to launch strategic bombing missions on key targets such as factories, trenches, ports, docks or any other target that presented itself. German Zeppelin airships conducted some of the first bombing raids over England and prompted retaliation. On 21 November 1914, three Avro 504s of the RNAS undertook the first aerial bombing raid in British history. The three aircraft flew 250 miles across enemy territory to attack the Zeppelin sheds at Friedrichshafen. The Avro 504s carried 20lb bombs on board, and in an incredible feat of navigation, they successfully found their target and caused extensive damage.

The Avro 504 proved to be a stable and reliable aircraft and as such made a very successful trainer in the years to come. However, it was not designed as a bomber. Although, it achieved its aims in the first bombing mission described above, it soon became vulnerable to enemy scouts and was therefore withdrawn as a bomber. The Avro 504 replica at Solent Sky Museum in Southampton has its side panels removed providing an interesting perspective. The mannequins in the display demonstrate how bombs would have been stored and dropped in very early raids.

Opposite: Airco DH9 – the world's only airworthy World War One bomber, owned by the Historic Aircraft Collection and pictured with a Hucks starter.

Strategic bombing was initially carried out from land or sea via the bombardment of shells, but range was clearly an issue. As soon as the aircraft was put into military service, it was considered a useful means to deliver ground attacks on targets previously out of reach. However, at the outbreak of the war there were no purpose-built bombers in service or on order for the RNAS or the RFC. Instead, the aircraft available were hastily adapted to carry out these missions. No. 3 Wing of the RNAS was the first unit specifically formed as a strategic bombing force. It was first equipped with Sopwith 1½ Strutters that were adapted into single-seat bombing variants. These hastily adapted bombers were also adopted by the French forces.

As the war progressed, many purpose-built bombers were designed and put into service. However, many of these types such as the Short Bomber, Airco DH4, Armstrong-Whitworth FK8, Handley Page Type O/100 and Type O/400 are now completely extinct. Owing to their size and complexity, very few World War One bomber replicas have been built, although TVAL have built an FE2b and there is a long-term project in the UK to recreate a Handley Page Type O/400.

This replica Sopwith 1½ Strutter at the RAF Museum Cosford was built to original Sopwith factory drawings and flown in 1980. It wears the markings of A8226, which was flown by C Flight of No. 45 Squadron of the RFC in April 1917. The Sopwith 1½ Strutter bomber variant could carry four 56lb bombs.

The Royal Aircraft Factory RE8 was designed as a reconnaissance aircraft to replace the floundering BE2. However, because of a lack of purpose-built machines, it was also adopted for bombing. No. 9 Squadron of the RFC took part in several bombing raids between June 1918 and the Armistice. RE8s were adapted to carry 20lb bombs on all flights so that they could attack any ground targets should the opportunity arise. The aircraft type was also used for night bombing and short-range tactical bombing raids on targets behind the lines. RE8 aircraft were also required to drop ammunition into the troops on the front line from the air.

The aircraft pictured here is a reproduction airframe built by TVAL in New Zealand. It is marked up as a No. 9 Squadron machine (A3930), which was originally built by Napier in London in 1917. It is currently on display in the Graham White building at the RAF Museum in Hendon.

Strategic bombing was undertaken by both sides during the war. The first aerial threat to Britain was from the German Zeppelins. They conducted waves of attacks on British cities between 1915 and 1916. Home defence squadrons were established at newly built aerodromes across the country to counter this threat. No. 37 Squadron based at Stow Maries Great War Aerodrome in Essex was vital to the protection of the south east of England. They initially operated BE2 aircraft as seen here conducting an engine run at the aerodrome, which has survived intact today.

From the spring of 1917, the German strategic bombing effort was intensified as they increasingly employed newly developed long-range bombers such as the Gotha G.V. Home defence efforts were doubled and aircraft such as the Avro 504 were adapted to perform these duties and were even employed as night fighters to counteract the increase in night-time raids. The aircraft pictured opposite is presented in the markings of No. 77 Squadron of the RAF based at East Lothian towards the end of the war. Although around 100 aircraft could easily be scrambled, home defence often lacked coordination in the air, and despite individual displays of gallantry, the impact on the German bomber force was limited.

The Royal Aircraft Factory FE2b was the third design to use the FE2 (Farman Experimental) tag, a designation given to pusher aircraft, with the engine behind the cockpit. This layout enabled a forward-firing machine gun prior to the invention of a synchroniser. The FE2b was initially powered by a 120hp Beardmore, although this was later upgraded to a 160hp model. It was operated as a fighter and a bomber and often carried the standard aerial camera of the time for reconnaissance duties too.

The observer sat in a precarious position right at the front of the aircraft, which was particularly hazardous during landings where any minor accident could prove fatal. The observer was armed with a .303 Lewis machine gun that was specially mounted to give a wider field of fire. Later models also gave the pilot an additional machine gun capable of firing behind the aircraft, which went some way to reducing the aircraft's vulnerabilities. As a fighter, the FE2b did much to counteract the Fokker Scourge in 1916. The FE2bs of No. 25 Squadron of the RFC even claimed victory over German flying ace Max Immelmann, although the Germans credited his loss to friendly fire.

Taking the Fight to the Enemy: Bombing Aircraft

Almost one third of the FE2bs built were produced as bombers. It was operated as both a day and night-time bomber and continued in service in the night role until August 1918. At its peak, the FE2b equipped 16 RFC squadrons on the Western Front and six home defence squadrons in Britain. Owing to the design limitations of the pusher layout, the aircraft was soon outclassed as a fighter, but it was generally well liked by its crews and could prove problematic to the enemy in the hands of the right crew.

The RAF Museum's FE2b is largely an original aircraft built in 1917, but it required many new components such as new wings, tailplane and undercarriage during its restoration in the late 1980s. It now appears as A6526. The original A6526 joined No. 148 Squadron on 2 April 1918, later flying with No. 58 Squadron before eventually joining No. 102 Squadron on 28 September. Here it was damaged on a night bombing sortie by ground fire. Although it made a forced landing, it was deemed beyond repair and did not serve again.

engines proved problematic in service, as they lacked the required power for carrying the heavy loads and were also plagued with reliability issues. The DH9 proved to be a less capable aircraft than the DH4 it was designed to replace. The arrival of the DH9a would rectify many of these issues.

The DH9 was the most produced aircraft of World War One; over 4,000 were built but very few have survived. Following the Armistice, many DH9s were surplus to British requirements and therefore shipped abroad to other nations as part of the Imperial Gift scheme. Of the six remaining DH9s, only two are viewable in the UK and this is only down to a chance find; left over from the Imperial Gift, two airframes remained largely forgotten in Bikaner, India. Guy Black of Aero Vintage (Retrotec) Ltd managed to negotiate the return of these airframes to England and undertook a lengthy restoration in partnership with the Imperial War Museum. The result was the return of one to the air whilst the other was prepared for static display at Duxford.

The Airco DH9 was developed in 1917 after it became clear that a bomber capable of carrying heavier loads was needed if strategic bombing was to have any impact. It was based around the structurally similar DH4 also designed by Geoffrey de Havilland, hence the DH designation. The first production machines were fitted with a Siddeley B.H.P. (Beardmore-Halford-Pullinder) engine, which was eventually replaced with a more powerful, lightweight variation known as the Siddeley Puma. Both

The static example, D-5649, was built in June 1918 and has been restored to non-flying condition using many of the original parts from the two recovered airframes. As this aircraft was not built to fly, it has been able to utilise authentic parts that would not be suitable for flight. Great care was taken to produce an authentic representation of the aircraft type, with a high standard of conservation employed.

Despite its shortcomings, the Airco DH9 saw extensive service in the last few years of the war. It was used in many bombing raids over Germany and the Western Front. It also saw service in anti-submarine patrols in the Middle East and the Indian sub-continent. After the war, it carried out military service all over the world and was widely adopted for civilian use. Its foundations in design led to the development of several light aircraft including the iconic de Havilland Tiger Moth.

E8894 was built in the spring of 1918 by the Aircraft Manufacturing Company, based at Hendon. It is not thought to have played an active role in the war but was selected for transport to India following the Armistice. From the two airframes recovered by Guy Black, it was selected as the most suitable for return to flight and although a significant number of parts have been rebuilt, it maintains a high level of authenticity. It has recently returned to the skies and is now the only genuine World War One bomber flying anywhere in the world.

The aircraft of World War One lacked the power, production numbers and accuracy to make any real impact on the outcome of the war via strategic bombing. However, the impact of bombing on the morale of the enemy should not be underestimated. The Great British public felt safe and remote from the impact of the war until the Zeppelins and bombers began to reach their cities. It should also be noted that the resources removed from the front line to make up home defence squadrons on both sides provided some relief to the opposition. The lessons learned in this conflict would influence future policy and aircraft designs eventually leading to considerably more significant air raids in World War Two and beyond.

CHAPTER 9
THE BIRTH OF THE ROYAL AIR FORCE

The Birth of the Royal Air Force

Although the RFC and RNAS had valiantly defended Britain, provided vital army reconnaissance and taken the fight to Germany, it soon became clear that a more centrally organised approach to air warfare was required. Increasingly devastating German attacks on London prompted a review into British air power. Lt Gen Jan Smuts was commissioned to conduct a review and in his six-page report, known as the 'Smuts Report', he recommended the formation of an independent air force on par in status with the British Army and Royal Navy.

Smuts suggested that the RFC and RNAS be amalgamated into the Royal Air Force (RAF). Prime Minister David Lloyd George concurred, and the RAF was officially formed on 1 April 1918. Britain was the first nation in the world to form its own stand-alone air force, which, on its formation, was equipped with 20,000 aircraft and over 300,000 men. This was the largest air force in the world at that time. A far cry from the four squadrons of ramshackle aircraft that had plodded their way across the English Channel four years earlier.

Opposite: The Great War Society recreates a World War One aviation scene at Stow Maries Great War Aerodrome.

Right: Stow Maries Great War Aerodrome's Sopwith Pup replica at sunrise.

Hugh Montague Trenchard is generally known as the 'Father of the Royal Air Force'. He was already head of the RFC in the field and was appointed as the first Chief of Air Staff for the RAF in April 1918. However, after heated disagreements with the then Air Minister Lord Rothermere, he soon resigned his commission. Before the month was out, Rothermere was forced to resign and the new Air Minster, Sir William Weir, was quick to appoint Trenchard as the General Officer Commanding. Trenchard was able to guide the newly formed RAF through the remainder of the war and then set about creating the foundations for the future, including setting up the RAF College at Cranwell: the world's first military air academy.

Despite the RAF being a new force, it initially relied on aircraft inherited from the RFC and RNAS, which included many aircraft that had long been out-dated by technological advancements. Aircraft such as the Sopwith Pup (pictured in the foreground) had been in service since 1916, and two years of war had seen rapid developments in aviation. More recent types such as the Royal Aircraft Factory SE5a (pictured in the background) were more capable but not available in the numbers required, forcing obsolete aircraft to continue in frontline roles that they were no longer suited for.

types were also addressed, and frontline fighters were brought back from France to help prioritise the defence of London.

Initially, the SE5 was chosen, which proved successful until the Gotha bombers switched to night-time raids. The SE5 had limited forward visibility, which made landings at night hazardous. The Wolseley Viper and earlier Hispano engines also took too long to warm up – delaying any possible interceptions. In the end, the Sopwith Camel proved a suitable replacement.

Shuttleworth Collection's Sopwith Camel is pictured here on the left next to its SE5a.

One of the first priorities of the Royal Air Force was to counteract the threat posed by the German Gotha bombers that were continuing to wreak havoc over British cities. The Sopwith Pup and BE2 were initially tasked with home defence duties, but these were outdated aircraft in 1918. One of the first improvements to home defence was the construction of an operations room in the basement of Hotel Cecil in London. Here, the incoming air raids could be plotted, and warnings could be given as part of a complex rapid response system. The aircraft

The first new aircraft type to enter service with the RAF was the Airco DH9a. Even this was a modification of the DH9, which found its origins in the DH4 that had been in service for some time. Despite great advances in aeroplane technology, Britain was still playing catch-up with engine production. In the early stages of the war, British aircraft were mostly powered by continental-built engines but more recently, superb engine designs had emerged from Bentley, Wolseley and Rolls-Royce. For the DH9a, Britain would turn to America for the 400hp Packard Liberty 12 engine to power its new bomber. The team at Westland Aircraft was already experienced in building the DH9 and took the lead in the conversion to the new engine. The result was a more powerful, reliable bomber that would serve with No. 110 Squadron of the RAF in France from 31 August 1918. After the war, the aircraft was so successful that it would continue in service until 1931.

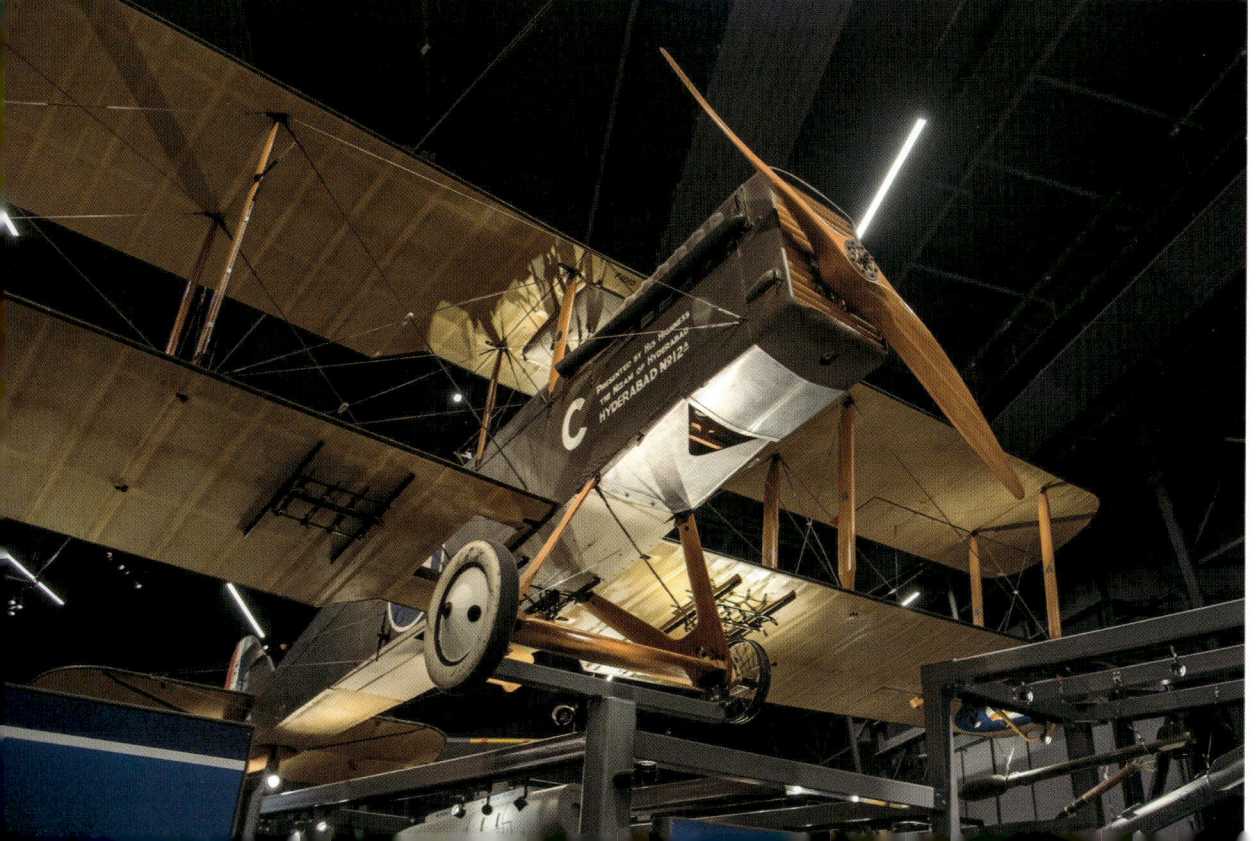

The RAF Museum's DH9a (F1010) was built by Westland Aircraft works at Yeovil in June 1918. It was one of the few to be built with an American 400hp Packard Liberty engine. It was delivered to No. 110 Squadron at Kenley in Surrey, which was uniquely funded by His Serene Highness, the Nizam of Hyderabad, hence the markings on the side of the aircraft. F1010 was then flown to France where it was often taken on bombing raids by its regular crew: pilot Capt Andrew Glover from Liverpool and observer Lt William George Lewis Badley from South Africa. Following a minor accident, the aircraft was adopted by Germany and put on display there until the beginning of World War Two. After several years based in Poland, the newly formed RAF Museum was able to acquire it for display in the UK where it has remained ever since.

The Sopwith Dolphin was the world's first multi-gun single-seater fighter. It first flew in May 1917 and entered service with the RFC at the beginning of 1918 before being passed on to the RAF in April. When the Dolphin first joined the fighting on the Western Front, it was frequently mistaken for a German aircraft and attacked by other Belgian and British pilots. Despite this inauspicious start, the aircraft performed well as a fighter. It was powered by a 200hp Hispano-Suiza 8B engine and featured an unusual staggered wing configuration with the lower wing forward of the upper wing. This layout affected novice pilots who found it hard to keep the aircraft level because of restricted views of the horizon. The views above, however, were much improved and the experienced pilot found this a major advantage in combat.

Many units were equipped with the Dolphin, but No. 87 Squadron was the most successful. It's crews shot down 89 enemy aircraft during the last nine months of the war. After the Armistice, the Dolphin was quickly phased out of service and did not continue as an RAF machine post-war. This Sopwith Dolphin is based at the RAF Museum in Hendon and is a reproduction aircraft that features many original parts including a fuel tank, radiator, bulkhead section, wheels, struts and tailfin. The aircraft carries the data plate from original Dolphin C3988, which is thought to have been used as a training aircraft in January 1918.

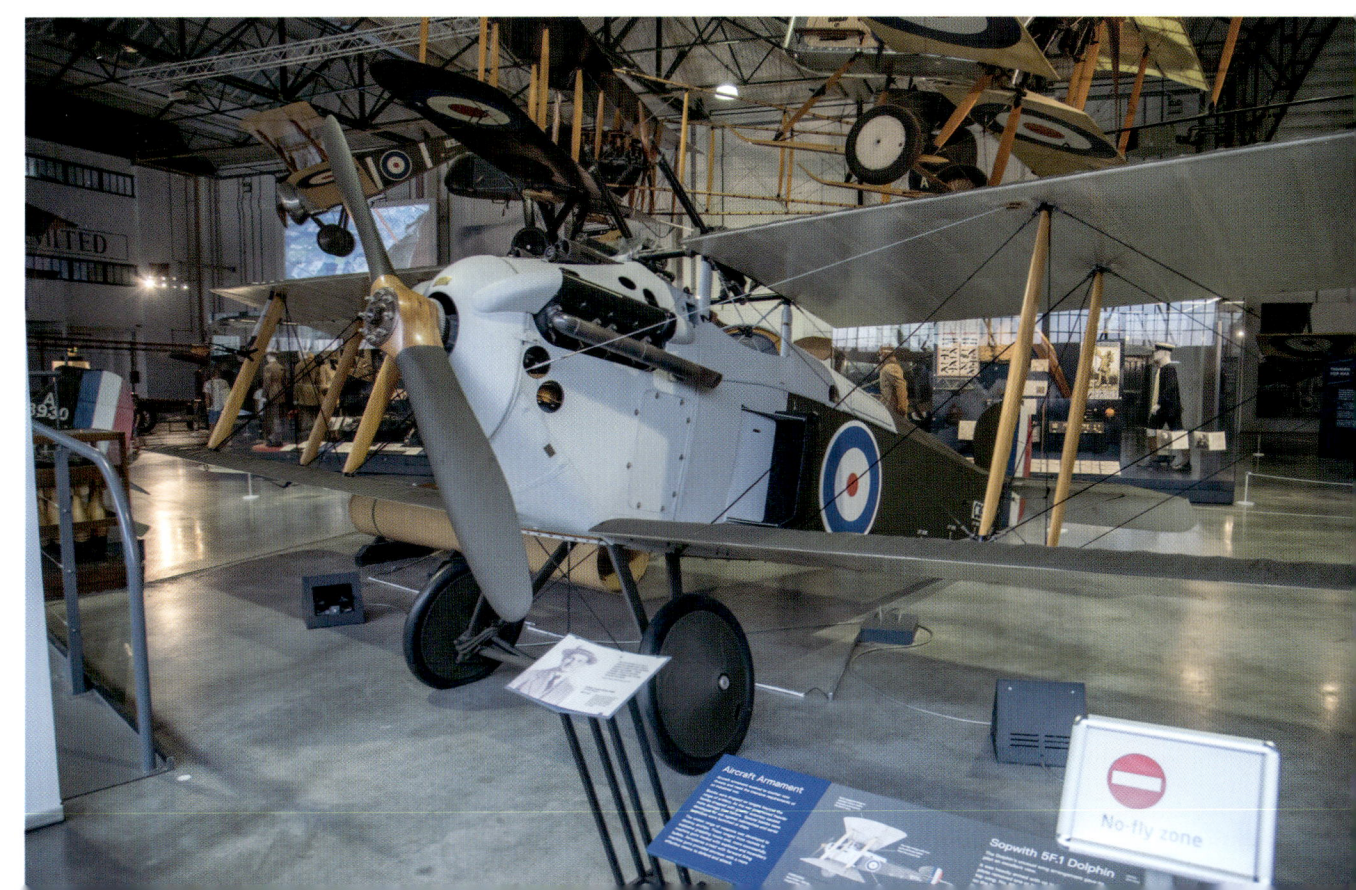

Innovations in aircraft design continued and soon the RAF would start to receive its own aircraft. Among the new generation was the superlative Sopwith Snipe, which was the latest in a line of aircraft from the Sopwith 'Zoo' tracing back from the Tabloid, Strutter, Pup and Camel. The Snipe was slightly smaller than the Camel, but the centre section of the upper wing was uncovered to give the pilot better visibility. It was also powered by a Bentley BR.2 rotary engine giving it an impressive performance and top speed of over 120mph. The new engine technology also supplied

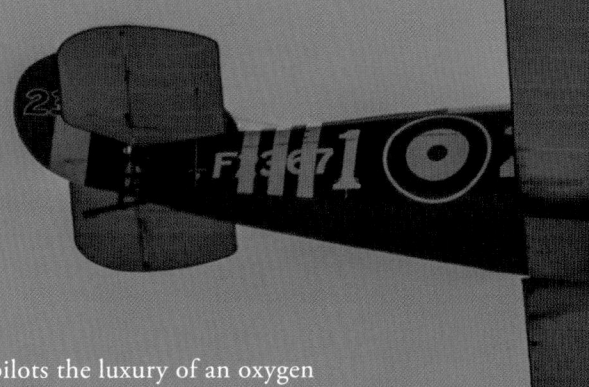

power for a heated cockpit and gave the pilots the luxury of an oxygen supply. The Snipe was used as both an all-out fighter and a ground-attack aircraft.

This replica Sopwith Snipe was built by TVAL in New Zealand. It is an exact reconstruction of Sopwith Snipe F2367 of No. 70 Squadron, which was on occupational duties in Germany immediately after the war. It is based in New Zealand but registered in the UK as G-CKBB. It was recently loaned to the WW1 Aviation Heritage Trust, making occasional appearances at UK airshows before returning to New Zealand in 2017.

The Sopwith Snipe first reached the Western Front with No. 43 Squadron of the RAF at the end of August 1918. In the final three months of the war, the Snipe proved itself to be the best Allied fighter to see action during the conflict. Although it was not fully deployed across the Western Front, pilots achieved great success in the Snipe. For example, Maj William G Barker was awarded the Victoria Cross for his single-handed engagement with 15 Fokker D.VIIs in October 1918. The Canadian pilot served most of his time on the Italian front but was given a ten-day pass on the Western Front where bad weather grounded him for nine days before he was finally unleashed in the Snipe. The result was four enemy aircraft destroyed by Barker before he was forced to make an emergency landing.

Fewer than 100 Snipes saw service during the war, as the hostilities were over before it could more widely equip other squadrons. Many of the wartime orders of the Snipe were cancelled but production did continue into 1919, by which time almost 2,000 had been built. Following the war, the Sopwith Snipe would see service in the Russian Civil War and remain on strength in the RAF until it was declared obsolete in 1928.

This Sopwith Snipe is a composite of various elements taken from other Snipes, including a significant number of parts from E6655. The aircraft was assembled by TVAL in New Zealand and arrived for display in the UK in 2012. E6655 was one of 150 Snipes ordered from Coventry Ordnance Works on 20 March 1918. It arrived with the RAF in 1919 and flew with No. 1 Squadron from RAF Hinaidi in Iraq in 1926. The Bentley BR.2 engine is an original and the tailplane was acquired from the Shuttleworth Collection, which had been using a Snipe tailplane on its Sopwith Pup. The rest of the parts were either newly rebuilt or salvaged from the RAF Museum's stores.

The end of the war brought with it many cuts in defence spending and an overall reduction in the aircraft and personnel required for the RAF. The future of the RAF itself was in doubt and the service had to wait nine months to have its status confirmed. However, it would be cut down to just 35,000 individuals, including 6,500 permanently commissioned officers. Fortunately, the world's first independent air force survived the cut and lived to fight another day. In 2018, the RAF celebrated its 100th birthday with several celebrations taking place, including a 100-aircraft strong flypast over London, which included 22 Eurofighter Typhoons forming the number 100 in the sky – a mark of how far military aviation has come since World War One.

CHAPTER 10
SUMMARY

Summary

Throughout the four years of World War One, aviation and its use by the military underwent dramatic changes. At the beginning of the war, powered flight was just 11 years old and the knowledge of aerodynamics, aircraft design and flight was still in its infancy. The combustion engine was also relatively new technology, and designers were constantly trying to develop lighter engines with more power. Many of the early aircraft were so underpowered that their top speed was barely above the prevailing wind they were flying into, making progress very slow indeed.

By the time the RFC and RNAS merged to form the world's first independent air arm, the RAF aeroplanes in service were unrecognisable from those flown at the beginning of the war. By 1918, aircraft were being used in all theatres of war. They were deployed in numerous roles on the Western Front, Italy, the Middle East and the Balkans. They were also employed as home defence in Britain and were widely required to operate from sea, either as floatplanes or as adapted carrier-borne aircraft. Despite the use of the aeroplane being so widely spread, its impact on the outcome of the war is often debated.

Opposite: The GWDT's BE2c simulates being shot down.

Right: The GWDT recreates a scene as would have been seen over the Western Front.

Fokker D.VII at Hendon.

During World War One, a plethora of aircraft types saw service on both sides; technology was advancing so quickly that both the Central Powers and the Entente had to keep pace with each other. As soon as a new aircraft type arrived on the front line, its opposition was suddenly obsolete. A constant arms race led to a mind-boggling number of new types entering service each year. Often, aircraft were rushed into service barely beyond the prototype stage. Unsurprisingly, most aircraft had a limited shelf life and types were often withdrawn or out-of-date within months of their arrival.

German aircraft technology was forced to move just as quickly as the Ententes, and, as such, their range of aircraft was equally as eclectic. The most successful types, such as the Fokker Dr I, Albatros D.Va and Fokker D.VII, proved formidable when introduced into the fight. As soon as a new type emerged, the enemy would immediately set about attempting to capture it for analysis and, therefore, many trends became apparent, including the Triplane craze, which was started by the British but widely adopted by the Germans. Equally as groundbreaking were the tactics employed by the Central Powers who often led the way in the use of squadrons and formations for aerial patrols.

Above: Fokker Dr I with DH9 in the background at Duxford.

Above: An LVG C VI undergoes conservation work at Cosford.

Below: Fokker Dr I of the GWDT.

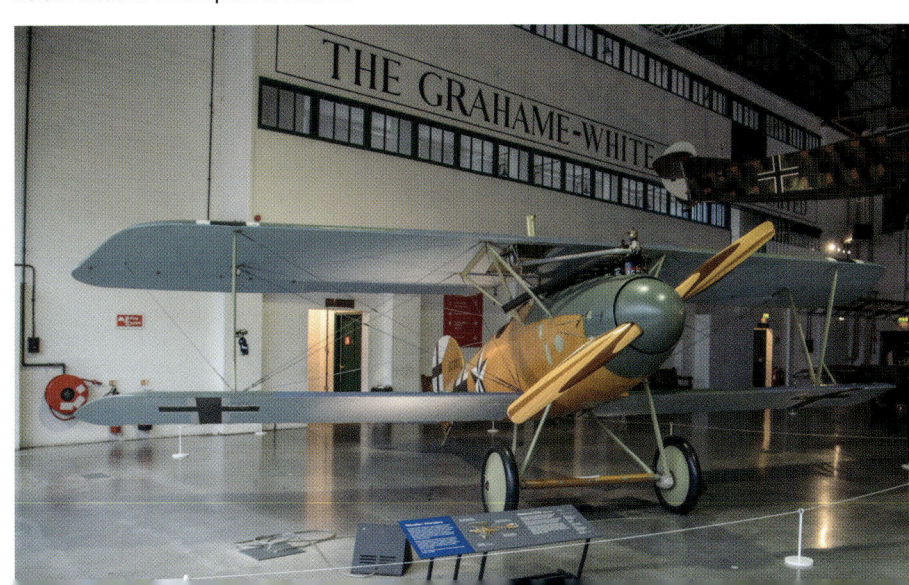

Below: Albatros D.Va replica at Hendon.

In the former part of the war, British aircraft technology lagged far behind most of its allies. In Europe, France had really embraced powered flight and was well prepared for taking the aeroplane into battle. Britain employed a wide range of French-built aircraft that helped to bolster its numbers and in many cases were far superior to those produced by British factories at the time. Even when British designs were dominating proceedings, they were often powered by French- or Spanish-built engines. Only at the very end of the war did Britain's aircraft industry find its feet and start producing well-thought-out airframes with reliable, powerful British engines.

French aircraft, such as the Cauldron G.3, were in service in France before the war but only taken on by Britain after the war broke out. This example was built in 1916 and is thought to have served with the Belgian Air Service during the war. It is now on display at the RAF Museum in Hendon and represents aircraft 3066 of the RNAS flying school at Vendôme as it would have appeared in 1917.

When the Americans entered the war in 1917, they were just as unprepared, even though powered flight was born in the USA. Throughout their involvement in the war, they made use of European designs such as the SPAD XIII, which was a French design also adopted by Britain. The SPAD XIII pictured here is a replica hanging from the ceiling at the American Air Museum at Duxford.

World War One pilots were brave beyond measure. Very few men had experience of flying in 1914, and so, pilots were often chosen for their skills on a horse or a sailing boat in the hope that skills of balance and proprioception would transfer to aircraft handling. The aircraft they flew were so primitive that they required constant management in the air. They were built from lightweight, flimsy, flammable materials that could fail at any time. The pilots were protected by little more than thin layers of canvas and were not allowed parachutes, as it was thought to inspire cowardice.

The life expectancy of an early World War One pilot was measured in weeks; he was not expected to achieve more than 60 flying hours before meeting his end. Many pilots were killed in accidents through inexperience, lack of instruction or mechanical failure before they even saw combat. It was not a job for the faint-hearted, but the heroics of the aces who became the 'Knights of the Skies' will be remembered forever. Although the war in the air between 1914 and 1918 was little more than a small fraction of the overall conflict, the lessons learned changed the shape of the military forever. Although this book has focused on the aircraft, we must always remember the human cost in the baptism of fire that was the development of military aviation. *Lest We Forget*.

Sopwith Pup replica and the Great War Society re-enactors at Stow Maries Great War Aerodrome.

BIBLIOGRAPHY

Black, Guy, *DH9: From Ruin to Restoration*, Grub Street (2019)

Bremner, David, *Bristol Scout 1264: Rebuilding Granddad's Aircraft*, Fonthill Media (2018)

Cotter, Jarrod, *Sopwith Camel: Owners' Workshop Manual*, Haynes (2016)

Ellis, Ken, *Wrecks & Relics*, 27th Edition, Crecy Publishing (2020)

Garton, Nick, *Royal Aircraft Factory S.E.5: Owners' Workshop Manual*, Haynes (2017)

Hamilton-Paterson, James, *Marked for Death: The First War in the Air*, Head of Zeus Ltd (2015)

Hare, Paul, R, *Mount of Aces: The Royal Aircraft Factory S.E.5a*, Fonthill Media (2013)

Herris, Jack and Pearson, Bob, *Aircraft of World War I*, Amber Books (2018)

Hoblyn, Ernie, *The Great War Display Team*, Amberley Publishing (2019)

Johnson, J E, *Full Circle: The Story of Air Fighting*, Cassell & Co (1964)

Rimell, Ray, *World War One Survivors*, Aston Publications (1990)

Simons, Graham, M, *De Havilland Enterprises: A History*, Pen & Sword Aviation (2016)

Mahoney, Ross (Ed.), *First World War in the Air*, Royal Air Force Museum (2015)